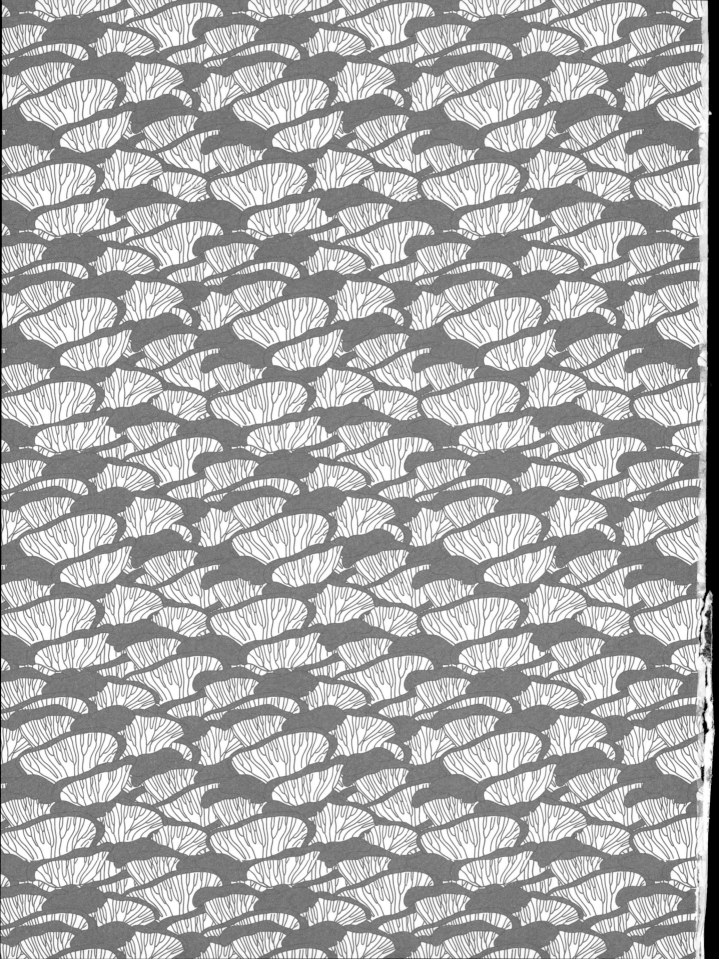

Fantastic Fungi

Fantastic Fungi

HOW MUSHROOMS CAN HEAL, SHIFT CONSCIOUSNESS & SAVE THE PLANET

EDITED AND INTRODUCTION BY **PAUL STAMETS**
AFTERWORD BY **LOUIE SCHWARTZBERG**

ESSAYS AND INTERVIEWS FROM
THE WORLD'S LEADING EXPERTS INCLUDING

MICHAEL POLLAN ❋ **ANDREW WEIL**
SUZANNE SIMARD ❋ **PAUL STAMETS** ❋ **ROLAND GRIFFITHS**
JAY HARMAN ❋ **WILLIAM RICHARDS** ❋ **EUGENIA BONE**
AND MANY MORE

EARTH AWARE

San Rafael, California

CONTENTS

INTRODUCTION • PAUL STAMETS 10

SECTION I • FOR THE PLANET 16

CHAPTER 1: MYCELIUM: THE SOURCE OF LIFE • **SUZANNE SIMARD** 18

    SPOTLIGHT ON: OLD-GROWTH FORESTS: CARBON WARRIORS • **PAUL STAMETS** 22

CHAPTER 2: NO STRAIGHT LINES • **JAY HARMAN** 26

    FUNGI FACTS: WEARABLE, BUILDABLE MYCELIUM 33

CHAPTER 3: THE WOOD WIDE WEB • **MERLIN SHELDRAKE** 34

    FUNGI FACTS: BILLIONS AND BILLIONS • **NIK MONEY** 37

CHAPTER 4: MUSHROOM MAD! • **GIULIANA FURCI** 40

CHAPTER 5: CULTIVATING MYCO-LITERACY • **PETER McCOY** 44

    SPOTLIGHT ON: FUNGI ACTIVISM 47

CHAPTER 6: FUNGI: A BEE'S BEST FRIEND • **STEVE SHEPPARD** 48

CHAPTER 7: MYCOREMEDIATION: GROWING PAINS AND OPPORTUNITIES • **DANIEL REYES** 54

    SPOTLIGHT ON: THE AMAZON MYCORENEWAL PROJECT 59

    SPOTLIGHT ON: FUNGAL CLEANERS 59

CHAPTER 8: THE MUSHROOM REVOLUTION IS HERE • **TRADD COTTER** 60

CHAPTER 9: THE SKY IS FALLING, BUT THERE'S A NET • **PAUL STAMETS** 64

SECTION II • FOR THE BODY 72

CHAPTER 10: MUSHROOMS: PHARMACOLOGICAL WONDERS • **ANDREW WEIL** 74

    SPOTLIGHT ON: A "TAIL" OF RECOVERY • **PAUL STAMETS** 78

    FUNGI FACTS: THE HEALTH BENEFITS AND VARIETIES OF MEDICINAL MUSHROOMS 79

CHAPTER 11: FUNGI AS FOOD AND MEDICINE FOR PLANTS (AND US) • **EUGENIA BONE** 80

    DELICIOUS MUSHROOM RECIPES 88

    FUNGI FACTS: THE FIRST CULTIVATED MUSHROOM 90

CHAPTER 12: A FORAGER'S HOMAGE • **GARY LINCOFF** 92

    SPOTLIGHT ON: GARY LINCOFF: MYCO-VISIONARY (1942–2018) 95

CHAPTER 13: A MUSHROOMING OF INTEREST • **BRITT BUNYARD** 96

    SPOTLIGHT ON: DESERT TRUFFLES • **ELINOAR SHAVIT** 98
    FUNGI FACTS: MUSHROOMS ACROSS TIME: AN ANCIENT HISTORY • **ELINOAR SHAVIT** 103

CHAPTER 14: GROWING YOUR OWN • **KRIS HOLSTROM** 104

    FUNGI FACTS: EATING WILD • **KATRINA BLAIR** 107

CHAPTER 15: MUSHROOMS TO THE PEOPLE • **WILLIAM PADILLA-BROWN** 108

CHAPTER 16: OUR GLOBAL IMMUNE SYSTEM • **PAUL STAMETS** 112

SECTION III • FOR THE SPIRIT 118

CHAPTER 17: THE RENAISSANCE OF PSYCHEDELIC THERAPY • **MICHAEL POLLAN** 120

CHAPTER 18: PSILOCYBIN IN A TEST TUBE • **NICHOLAS V. COZZI** 124

CHAPTER 19: DOORWAYS TO TRANSCENDENCE • **ROLAND GRIFFITHS** 128

    SPOTLIGHT ON: TOUCHING THE SACRED • **ALEX GREY** 132

CHAPTER 20: A GUIDE THROUGH THE MAZE • **MARY COSIMANO** 136

    SPOTLIGHT ON: JOHNS HOPKINS SESSION TRANSCRIPTS 138
    SPOTLIGHT ON: SACRED CEREMONY • **ADELE GETTY** 140

CHAPTER 21: A GOOD DEATH • **STEPHEN ROSS** 142

    SPOTLIGHT ON: PSILOCYBIN RESEARCH: A CLINICAL RESURRECTION • **CHARLES GROB** 147

CHAPTER 22: THE MYSTERIES OF SELF-NESS • **FRANZ VOLLENWEIDER** 148

    SPOTLIGHT ON: MICRODOSING: REAL OR PLACEBO? 151

CHAPTER 23: OF APES AND MEN • **DENNIS McKENNA** 152

    COUNTERPOINT: THE STONED APE THEORY—NOT • **ANDREW WEIL** 155

CHAPTER 24: MYSTICAL EXPERIENCES ACROSS RELIGIOUS TRADITIONS •
ANTHONY BOSSIS, ROBERT JESSE, AND WILLIAM RICHARDS 156

    SPOTLIGHT ON: COUNCIL ON SPIRITUAL PRACTICES/HEFFTER RESEARCH INSTITUTE 159
    SPOTLIGHT ON: MEDITATION AND PSILOCYBIN • **VANJA PALMERS** 160

CHAPTER 25: CHANGING THE GAME • **PAUL STAMETS** 162

AFTERWORD • **LOUIE SCHWARTZBERG** 174

ACKNOWLEDGMENTS • **LOUIE SCHWARTZBERG** 176

INDEX 178

INTRODUCTION

PAUL STAMETS

PAUL STAMETS is the preeminent mycologist in the United States. He has discovered several new species of mushrooms, pioneered countless new techniques, published several best-selling books, and won numerous awards.

Mushrooms are mysterious.

They come out of nowhere suddenly, with their splendid forms and colors, and just as quickly, go away. Mushrooms' startling appearances and enigmatic disappearances have made them forbidden fruits for thousands of years. Only a few of the cognoscente—the shamans, the witches, the priests, and the wise herbalists—have gained a glimmer of the knowledge mushrooms possess.

Why?

It is natural to fear what is powerful yet unknown. Some mushrooms can kill you. Some can heal you. Many can feed you. A few can send you on a spiritual journey. Their sudden rise and retreat back into the underground of nature make them difficult to study. We have longer periods of contact with animals and plants and we usually know which ones can help or hurt us. Mushrooms are not like that. They slip into our landscape and exit shortly thereafter. The memory fades quickly, and we wonder what we saw.

Mushrooms are the fruit bodies of a nearly invisible network of mycelium, the cellular fabric beneath each footstep we take on the ground. Reach down and move a stick or a log, and you will see a vast array of fuzzy, cobwebby cells emanating everywhere. That's mycelium, the network of fungal cells that permeates all landscapes. It is the foundation of the food web. It holds all life together. Yet these vast underground networks, which can achieve the largest masses of any organism in the world and can cover thousands of acres, hide in plain sight; silent but sentient and always working tirelessly to create the soils that sustain life.

Over thousands of years we have accumulated a large body of knowledge when it comes to edibles. Starvation is a good motivator for finding novel foods. Our ancestors quickly learned that some mushrooms are not only nutritious but delicious. Mushrooms provide protein and vitamins, and they can strengthen our immune systems. They have been critical in our species' struggle for survival.

Many elderly people share joyous memories of going with their parents and grandparents on family trips into the forest to pick mushrooms. They have experienced that eureka moment of discovery and understand the challenges of identifying edibles and the danger of misidentifying toxic species. They know the reward and joy of a delicious meal foraged by their family from the natural world around them. All this can create meaningful memories that bond families across generations. Many mushroom patches are kept as family secrets, only shared with future generations. This is what

PAGES 2-3 *Sulfur tuft (*Laetiporus conifericola*)*. PAGE 4 *Unidentified mushroom species*. PAGE 6 Gymnopus *sp*.
PAGE 9 Coprinellus *sp*. OPPOSITE FROM TOP LEFT TO BOTTOM RIGHT *Mushroom species unknown;* Clathrus *sp*. *(© Taylor Lockwood);* Hygrocybe conica; Clathrus *sp*. *(© Taylor Lockwood);* Fistulina hepatica; Omphalotus olearius; *Mushroom species unknown;* Microstoma protractum; *Chanterelle;* Laetiporus sulphureus; *Mushroom species unknown;* Aleuria aurantia.

the mushroom experience does—it grows on you. It is like a mycelial thread through time, a bridge from our ancestors to us and to our descendants in the future.

It is the multiplicity of benefits, I think, that makes mushrooms so attractive to those who learn their uses. One theme that pervades indigenous cultures: It is the substances that are utilitarian, that help humans survive, that are threaded into the cultural fabric.

Though much of our ancestral knowledge about mushrooms has been lost to history, a lot of knowledge is being scientifically validated as we begin to study fungi in clinical contexts. Penicillin, made from *Penicillium*, began the era of antibiotics and has saved millions of lives. An endophytic fungus, *Taxomyces andreanae* was discovered to synthesize taxol, which can treat certain types of cancers. I personally discovered that extracts from *Fomitopsis officinalis* protect against the family of viruses that includes smallpox. Fungi often have antimicrobial properties, they can support the immune system, and they can prevent or heal viral diseases. They can do so much, and yet, we have only really begun to discover the endless possibilities that the fungal world holds when it comes to improving human health.

Many edible mushrooms are both delicious and good for you. However, most mushrooms, though they're not poisonous, do not taste good. What is deemed inedible by one culture is sometimes a delicacy in another. The poisonous *Amanita muscaria* is called fly agaric. Long before the invention of window screens, pre-Europeans chopped up fly agaric mushrooms and placed the pieces in bowls of souring milk on windowsills to attract and kill flies.

Not a good mushroom to consume? You may think that, but in Asia and elsewhere, foragers discovered that if you boil fly agaric mushrooms in water and rinse them three or more times, the water-soluble toxins are removed, thus rendering the mushrooms edible without ill effects. The berserkers of Viking lore reportedly consumed this mushroom before battle, and in the ensuing frenzy, they became quasi-robotic killers, as the mushrooms can induce uncontrollable repetitive motions and allow the person to ignore pain.

In Siberia, shamans would ingest *Amanita muscaria*. They eventually discovered that reindeer would consume their pee by eating the yellow snow. The Siberians used that knowledge to lasso and corral the stoned reindeer with ease. It's incredible that just one variety of mushroom can poison flies, herd reindeer, weaponize humans, and feed people (if prepared properly).

An ingredient in the evolution of cultures from ancient Europe to North America was the discovery of "magic mushrooms," particularly those containing psilocybin. Though they've been used in various contexts throughout human history, recent clinical studies in the United States and Europe show how doses of psilocin (the active ingredient in psilocybin mushrooms) help victims of trauma and terminal patients in fear of death and are even associated with a reduction in criminal tendencies.

Magic mushrooms have been ingested for millennia in Europe and Mexico. Preserving psilocybin mushrooms in honey is a practice in Mexico to this day. Before the Bavarian Beer Purity Act Law (*Reinheitsgebot*) in 1516, mushrooms had been used to spike beer. These hallucinogenic brews were part of local nature-worship practices. Some archaeobotanists suspect magic mushroom meads,

OPPOSITE TOP LEFT TO BOTTOM RIGHT Mucronella *sp. (© Taylor Lockwood);* Amanita muscaria; *Mushroom species unknown;* Cookeina tricholoma.

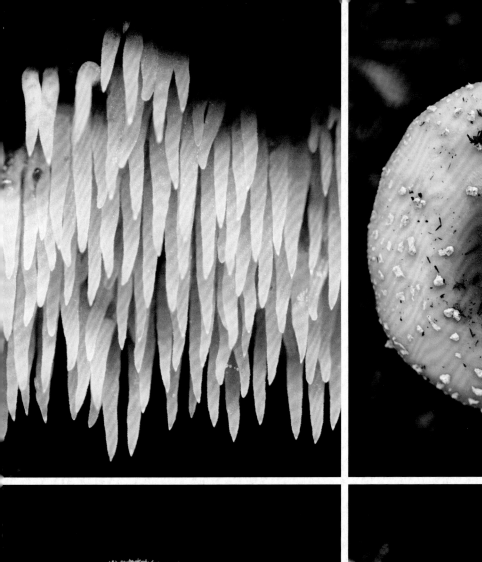

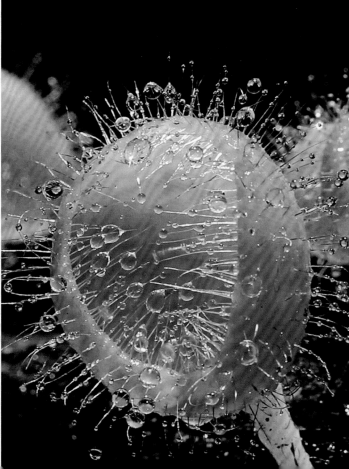

fermented honey-based brews infused with magic mushrooms, figured in early European and Mesoamerican rituals.

Honey, bees, and mushrooms are intimately connected. Let me explain. *Fantastic Fungi*, the book and the movie, have been a collaboration between filmmaker Louie Schwartzberg and myself for more than a decade. A few years ago, Louie completed a film on pollinators, *Wings of Life*, which includes bats, butterflies, and bees. After watching bees struggle, he grieved at seeing the greatest pollinators of all die off in massive numbers. He asked me a poignant question: "Paul, is there anything you can do to help the bees?"

Louie knew of my prior work with fungi and insects, but his question made me remember a bizarre event I'd witnessed in my garden. Back in 1984, I had two beehives. One July morning I went to water my mushroom patch. I noticed a cluster of twenty or so bees busy on the surface of my wood chips. Upon closer examination, I saw they were sipping on tiny droplets exuding from white threads of mycelium in between the woodchips. A continuous stream of bees traversed several hundred feet, from my beehives to my garden, from dawn to dusk for forty days. The bees moved the chips—giant wooden monoliths compared to their bodies—to uncover more succulent droplet-oozing mycelium underneath. Daily, they "milked" the mycelium.

Years later, after that conversation with Louie, my memory of these observations came back in a waking dream. That morning, I began connecting the dots between the bees in my garden and my work with the Department of Defense BioDefense program. That post-9/11 project led to the discovery that several of our polypore mushroom extracts were potent at reducing the effects of potentially weaponizable viruses, such as smallpox and influenza. I wondered if these same extracts could help bees fight deadly viruses that Varroa mites inject into them, a major cause of bee colony collapse.

Upon waking, I knew what I had to do: I needed to test those extracts on bees to see if I could help them fend off debilitating viruses.

Four years later, our team, working with Washington State University and the United States Department of Agriculture, found that the extracts of mycelium of woodland polypore mushrooms reduce the bee-killing viruses. Once the bees sip our extracts, bee viral loads plummet thousands of times. Bee lifespans are extended. This paradigm-shifting technique could help us assist bees to overcome colony collapse disorder and strengthen worldwide food biosecurity. A happenstance thought-dance between an artist and a scientist spawned this historic breakthrough.

Mushrooms are food for the body and medicine for the soul. The *Fantastic Fungi* film and this book are your portals to a grander wonder. In these pages, we'll explore studies pointing to mycelium as a solution to our gravest environmental challenges, examine research that reveals mushrooms as a viable alternative to Western pharmacology, and learn about fungi's marvelous proven ability to shift consciousness. Welcome to the mushroom underground. We are all connected!

OPPOSITE TOP Tremella fuciformis *(© Taylor Lockwood).* OPPOSITE BOTTOM LEFT *Mycelium.*
OPPOSITE BOTTOM RIGHT *Mushroom species unknown.*

SECTION I
FOR THE PLANET

There is a feeling in this world, the pulse of eternal knowledge.

When you sense the oneness, you are with us.

We brought life to Earth.

You can't see us, but we flourish all around you.

Everywhere, in everything, and even inside you . . .

from your first breath, to your last . . .

in darkness, and in the light.

We are the oldest, and youngest.

We are the largest, and smallest.

We are the wisdom of a billion years.

We are creation.

We are resurrection.

We are condemnation and regeneration.

The "third kingdom" of fungi and mushrooms is a realm of mystery on whose secrets the future of life on Earth may depend. At a time when solutions to our planet's most pressing challenges seem as elusive as ever, the ground beneath our feet may hold the most promising answers.

Scientists and researchers around the world, trained in universities and inspired by their passion for the planet, are discovering that underground networks of interconnected organisms are revealing a new story about the planet's ability to heal itself. The innate intelligence of these networks—the result of billions of years of evolution—has much to teach us.

In this section, experts explore how humans can utilize mushrooms to restore our planet.

MYCELIUM: THE SOURCE OF LIFE • SUZANNE SIMARD *(pg.18)*

SPOTLIGHT ON: OLD-GROWTH FORESTS: CARBON WARRIORS • PAUL STAMETS

NO STRAIGHT LINES • JAY HARMAN *(pg.26)*

FUNGI FACTS: WEARABLE, BUILDABLE MYCELIUM

THE WOOD WIDE WEB • MERLIN SHELDRAKE *(pg.34)*

FUNGI FACTS: BILLIONS AND BILLIONS OF SPORES • NIK MONEY

MUSHROOM MAD! • GIULIANA FURCI *(pg.40)*

CULTIVATING MYCO-LITERACY • PETER McCOY *(pg.44)*

SPOTLIGHT ON: FUNGI ACTIVISM

FUNGI: A BEE'S BEST FRIEND • STEVE SHEPPARD *(pg.48)*

MYCOREMEDIATION: GROWING PAINS AND OPPORTUNITIES • DANIEL REYES *(pg.54)*

SPOTLIGHT ON: THE AMAZON MYCORENEWAL PROJECT

SPOTLIGHT ON: FUNGAL CLEANERS

THE MUSHROOM REVOLUTION IS HERE • TRADD COTTER *(pg.60)*

THE SKY IS FALLING, BUT THERE'S A NET • PAUL STAMETS *(pg.64)*

CHAPTER 1
MYCELIUM: THE SOURCE OF LIFE

༄ ༄ ༄

SUZANNE SIMARD

SUZANNE SIMARD is a professor of forest ecology at the University of
British Columbia's Department of Forest and Conservation Sciences in Vancouver.

What goes on beneath a forest floor is just as interesting—and just as important—as what goes on above it. A vibrant network of nearly microscopic threads is recycling air, soil, and water in a continuous cycle of balance and replenishment. Survival depends not on the fittest, but on the collective.

Imagine a log that was once was a tree. Maybe it died of old age or became infected by a disease and fell over. When it did, fungi spread into the log from the earth below and started decomposing it. These fungi are part of a vast network of underground vegetation called mycelium, composed of very tiny, cobweb-like threads of organic life called hyphae. All along the thousands of miles of mycelium occupying this one big log, the mycelium uses the fungi to send out enzymes and organic acids that break down the lignin in the wood's cell walls that gives the wood its structure and strength.

In the process of decaying, the wood releases its nutrients. Those nutrients become available to other organisms in the food web, including the fungus, which distributes the nutrients through the mycelium. After many challenges, these networks come to the surface and form mushrooms, the reproductive structures of fungi. Mushrooms are literally "the tip of the iceberg." When we see mushrooms, there's actually a vast network of mycelium hidden in the ground beneath them. Only about 10 percent of all fungi produce mushrooms. But when you pick a mushroom, you stand upon a vast, hidden network of fungal mycelium that literally extends underneath every footstep you take. These networks are the foundation of life. They create the soils that nourish all life on land. Without fungi, we do not have soil. Without soil, there is no life.

Without this metamorphic process, the planet would choke. Forests would be miles deep in organic decay. The only reason we can walk around in most woods is because thousands of species of fungi are decaying all of the organic detritus on the forest floor, recycling the dead material and beginning the renewal of life.

Once the log starts to decay and release nutrients, other organisms such as mites and nematodes move in and start eating the fungus, small bits of leftover wood, and other organic material, which they ultimately excrete. Some of that ends up in the nutrient cycle, and then other critters like springtails or spiders come along and eat the nematodes, and then other creatures eat the springtails and spiders, and so it goes up the food chain to larger and larger organisms. Even mushrooms become food for squirrels, and eventually something will eat the squirrel—maybe a bird, maybe a bear—and it's all

OPPOSITE TOP Leucocoprinus cepistipes. **OPPOSITE MIDDLE** Hemitrichia sp. (© Taylor Lockwood). **OPPOSITE BOTTOM** Mycelium.

linked back to that original wood-decomposing fungus. When each organism reaches the end of its life, it returns to the soil and continues replenishing the cycle.

In all ecosystems, death and decay are the fundamental beginning of life. If a forest never went through this process, it couldn't regenerate. It would crowd near big trees, leaving few gaps for young ones. The big trees would suck up the light, water, and nutrients. Decay organisms like fungi are crucial for that process of regeneration. They are the building blocks of the ecosystem, the fundamental starting place for how a forest grows.

MOTHER TREES AND THEIR FUNGAL PARTNERS

Forests are incredibly complex places with trees of many different sizes and compositions, depending on the type of forest. There are little trees coming up in the understory, and then there are the parents, the big trees that provide the seeds for the forest's continued diversity and generational health. We call the biggest and the oldest of these trees Mother Trees because they are connected below ground to all the other trees around them—their community—by what we call a mycorrhizal fungus.

A mycorrhizal fungus is a special kind of organism that forms a symbiotic relationship with the tree. It wraps its fungal body around soil particles, extracting nutrients and water that it then brings to the roots of the tree. In return for this precious nourishment, the tree obliges the fungus by providing it with the sugars that the fungus needs to survive. These sugars are infused with carbon that the tree has accumulated through photosynthesis.

Climate change is the result of an accumulation of greenhouse gasses in the atmosphere, and carbon dioxide (CO_2) is the biggest culprit. Carbon dioxide is also what plants photosynthesize. They put that carbon in different places, such as their leaves and trunks, but we now know that 70 percent or more of that carbon actually ends up below ground.[1] It's first stored in the cell walls of plants until it's traded for nutrients via these root exchanges with mycorrhizal fungi. Once the carbon has been absorbed by these fungi, it can stay underground for thousands of years. And surprisingly, when mycelium dies, it also locks in the carbon for extended periods of time, building a carbon reserve for the future.

These mycorrhizal fungi form a network of threads that bond with the roots of other trees in the neighborhood and connect them all, no matter the species, like underground telephone wires. The biggest and oldest trees—the Mother Trees—have the largest root systems and the most root tips intertwined with these fungi, and they therefore connect with more trees.

We often think of kin selection or kin recognition as an animal behavior, but our research is showing that these Mother Trees also recognize their own kin—their seedling offspring—through these mycorrhizal networks and communicate with each other through carbon.[2] Carbon is their universal language. The stronger trees support the weaker ones by regulating the flow of carbon between them. If a Mother Tree knows there are pests around and her offspring is in danger, she'll increase their competitive environment. A good example is the *Leucaena leucocephala* (commonly known as the white leaf tree or river tamarind, among other names) that is native to southern Mexico and northern Central America. When it senses a competing species, it releases a chemical into the soil that stops the competitor's growth. It's a magical thing and it could not happen without the fungi.

Scientists have been obsessed with competition and the survival of the fittest ever since Charles Darwin posited the theory of natural selection. In a lot of ways, our innate competitive instincts influence how we view each other and manage the natural world around us. We tend to favor those individuals who are big and strong and best positioned to win. We do this in forestry, in agriculture, and in fisheries. And yet the way people behave and thrive depends on a community. It's no different in the forest. The cooperation that goes on between trees and plants is just as important as the competition, which allows them to work together to adjust to changes and threats in their environment.

> "By maintaining and protecting the plants, the forests, and the fungal networks, we will help to support a beautiful, resilient community that is also a natural engine for restoring and maintaining a sustainable planet."

We've always thought of plants and trees and fungi as essentially inert objects that don't interact with each other, that don't build things. But my work and others' is showing that they need each other to share the load and grow as one: You do this, I'll do that, and together we will thrive. This gives me incredible hope. Nature wants to heal itself; even if you try to make a bare space, plants will fill it in. They want to be there. They just show up and start doing their thing. So by maintaining and protecting the plants, the forests, and the fungal networks, we will help to support a beautiful, resilient community that is also a natural engine for restoring and maintaining a sustainable planet.

NOTES

1. T. A. Ontl and L. A. Schulte, "Soil Carbon Storage," *Nature Education Knowledge* 3, no. 10 (2012): 35, https://www.nature.com/scitable/knowledge/library/soil-carbon-storage-84223790.
2. Monika A. Gorzelak, Amanda K. Asay, Brian J. Pickles, and Suzanne W. Simard. "Inter-Plant Communication through Mycorrhizal Networks Mediates Complex Adaptive Behaviour in Plant Communities," *AoB PLANTS* 7, no. 1 (January 2015), https://academic.oup.com/aobpla/article/doi/10.1093/aobpla/plv050/201398.

TOP *An unidentified mushroom and mycelium on tree bark.*

SPOTLIGHT ON:
OLD-GROWTH FORESTS: CARBON WARRIORS

I was born on July 17, 1955. At the time, the carbon dioxide levels on this planet were 310 to 320 parts per million. Now, we're over 400 parts per million, higher than at any point in at least the past eight hundred thousand years.[1] That's over a 20 percent increase in my lifetime. Such geo-atmospheric cycles usually take thousands or even millions of years to occur, and we're seeing it within my sixty-plus years of life. In terms of geologic cycles, this is sudden; nearly instantaneous. Scientists are sounding the alarms while denialists have their heads in the sand. This is an all-hands-on-deck moment.

The science has clearly established that mycelium networks act as the most significant biological storage sink for carbon—far more than what is stored in aboveground flora and trees.[2] But we need both, and old-growth forests are the best aboveground storage of carbon anywhere in the world. It used to be that the metric we used to measure the value of an old-growth forest was the economic value of the lumber. But that metric is quickly being offset with the advances in science.

We're going down a slippery slope. As we deforest the planet and cut down old-growth forests, we accelerate carbon loss, temperature rise, and ecosphere devastation. Biodiversity plummets. Humans experience poverty, disease, and unsustainability. This has all been well demonstrated.[3] Intact forest ecosystems, by comparison, provide more ecological services than just board feet of lumber. They clean the water, provide shade, and give communities plants, insects, and animals. Protecting our forests is essential not only for our survival now, but also for the survival of generations to come.

—PAUL STAMETS

OPPOSITE TOP LEFT AND RIGHT *An edible, parasitic mushroom usually found in Patagonia, southern Chile, and Argentina* (Cyttaria harioti).

OPPOSITE BOTTOM *Turkey tail* (Trametes versicolor).

LEFT *Schizophyllum commune.*

FOLLOWING PAGES Cookeina sulcipes.

NOTES

1. Rebecca Lindsey, "Climate Change: Atmospheric Carbon Dioxide," Climate.gov, accessed August 1, 2018, https://www.climate.gov/news-features/understanding-climate/climate-change-atmospheric-carbon-dioxide.
2. K. E. Clemmensen et al., "Roots and Associated Fungi Drive Long-Term Carbon Sequestration in Boreal Forest," *Science* (March 29, 2013): 1615-18; T. A. Ontl and L. A. Schulte, "Soil Carbon Storage," *Nature Education Knowledge* (2012): 35.
3. Luis Carrasco et al. "Unsustainable Development Pathways Caused by Tropical Deforestation," *Science Advances* 3, no. 7 (July 12, 2017).

CHAPTER 2

NO STRAIGHT LINES

∾ ∾ ∾

JAY HARMAN

JAY HARMAN is CEO and chief inventor for PAX Scientific, an AAAS-Lemelson Invention Ambassador, and an adjunct professor at Curtin University in Perth, Australia.

Nature has kept us alive since the beginning of human life. Its inherent creativity and innovation are now showing us how to build a truly sustainable society.

If you think about it, nature—life—has been around for a very long time, nearly four billion years. Most species that have ever existed are now extinct—99.999 percent of them, to be exact. A lot of evidence about who they were and what they did and how they were designed is still in the fossil record. And today, we have at least three million species of fungi on Earth. Some researchers believe there may be another one hundred million or more still undiscovered.[1] Every one of those species has hundreds, possibly thousands, of solutions to the problems that we humans, in our highly engineered world, are trying to deal with on a daily basis.

Biomimicry is the art and science of solving complex human problems by taking inspiration—and lessons—from nature. Humans have been doing this for at least a million years, from the aboriginals making boomerangs, which are wonderful copies of bird wings, to the way we build pottery modeled on how a potter wasp builds a nest, to many other applications. And so it was until the start of the Industrial Revolution. Now "mycomimicry" can be the template of creativity for launching new solutions based on the cleverness of these fungal networks.

Before assembly-line manufacturing took over, humans never built things in straight lines, except maybe buildings. Even ships were built intuitively, taking on the proportions and shapes of fish and sea birds. They were extremely strong and effective at getting the job done. The beginning of the Industrial Revolution, however, saw the arrival of explosive forces like piston-driven steam engines, and everything started favoring straight lines. For the first time in history, people were discovering—and applying—the laws of Newtonian physics.

It made a lot of sense. It made things easier. Mass production became much simpler and increasingly more important, along with the use of steam and, later, internal combustion. As humanity entered this paradigm, it started engineering tools to match that approach: flat, straight, and having nothing to do with nature. If we needed more energy, we didn't worry about perfect efficiency, which is fundamental to how nature works—it always uses the least to get the most done—we simply added more fuel. That worked really well for quite a long time until that energy inefficiency, along with a growing scarcity of materials, started creating some very negative consequences, multiplied today by billions of people.

OPPOSITE Aloe polyphylla.

✣ SOLUTIONS SOURCED FROM NATURE ✣

From nature's point of view, there is no such thing as a straight line and no such thing as an energy shortage. There never has been, never will be. Everything in our universe is made of energy. Even solid objects. We all know this from high school. Every atom is made of little fields of energy that happen to be in the shape of whirlpools which, it turns out, is the only shape that nature uses, from the biggest galaxy down to subatomic particles. Same geometry, same kind of spin. And when I realized this, I thought, "If a fundamental mechanism in nature is a whirlpool, I want to understand that." And so I spent quite a few years trying to capture the actual geometry of a whirlpool, and when I finally did, I put that knowledge to use. I created a simple little device that keeps the water in municipal water tanks—some as big as ten million gallons—moving to prevent stagnation and bacterial buildup.

There are endless such examples of nature providing a solution. Take, for instance, the damage of exposure to ultraviolet rays. Nonmelanoma skin cancer is the most common cancer in the United States, so it's vitally important to use sunscreen.[2] There are about 1,800 varieties of sunscreen on the market, but most of them aren't very good. They block out either UVA or UVB—but not both—and they all have chemicals that you rub into your skin. Many of them, researchers are now saying, are more dangerous to your health than actual exposure to the sun.[3]

"In May 2018, the state of Hawaii passed a law banning the sale or distribution of over-the-counter sunscreens containing oxybenzone and octinoxate. A study by Haereticus Environmental Laboratory, a nonprofit scientific organization, found that the chemicals harmed exposed coral, causing bleaching, deformities, DNA damage, and ultimately death."

—JAY HARMAN

So a few folks got to thinking: Nature is full of species dealing with ultraviolet exposure that don't suffer from skin disease. What are their strategies? One of them is particularly interesting: hippopotamus sweat. It's waterproof, antiseptic, antifungal, and antiparasitic, and it cuts out UVA and UVB. You put a drop of it on your hand and it spreads by itself. Researchers are now reverse-engineering those molecules in a way that will probably rewrite the book on sunscreens and completely disrupt that industry.

And here's another: We're all familiar with spiderwebs, five times stronger than the best steel in the world when you look at strength-to-weight ratios. One interesting thing with spiderwebs is that birds don't fly into them. Why is that? It turns out that nature has a strategy for keeping them separate: The spider weaves tiny UV reflectors throughout its web. We can't see them, but a bird can. A company in Germany has created a product called Ornilux that mimics that technology by weaving little UV reflectors through glass sheets. An estimated one billion birds die each year from hitting glass windows. It's already been shown that these little reflectors can reduce those bird strikes by more than 70 percent. We can save seven hundred million birds a year by implementing that technology.

ABOVE Gyrodon merulioides (© Taylor Lockwood). OPPOSITE BOTTOM *The pattern of water upon a sandy beach.*

❧ THE BIOMIMETIC BOTTOM LINE ❧

Since the start of the Industrial Revolution, we've been on a track where sustainability hasn't been an issue, hasn't been part of our consciousness. When I started with the Fish and Wildlife Department in the late '60s, I had never heard of the words *pollution* or *sustainability*. Fortunately, awareness is expanding now because we can no longer ignore what's happening. I talk to kids in schools and universities all over the world, and there's this extraordinary pessimism because they know that the future of this planet doesn't look good.

"We're looking at a revolution, a technological change, an about-face of our trajectory like nothing we've seen before."

And yet despite these forbidding potentials, we have remarkable capabilities that allow us to focus on alternative paths forward. There's a rapidly growing recognition that nature can teach us all the solutions we need to restore sustainability and create a viable future for the planet and its species—including humans. We've reached a confluence of both amazing tools and the ability to understand nature and reverse engineer its systems.

There are more college graduates today than at any time in history, and young people want something more. Wherever I give a talk about these possibilities, the kids come alive. They suddenly see that they aren't sentenced to death, that there is all this opportunity, and that they can actually reinvent their world. We're looking at a revolution, a technological change, an about-face of our trajectory like nothing we've seen before. A massive fundamental shift. I have no doubt it will happen. It *is* happening. And here's why.

In the twenty-first century, every decision comes down to the bottom line because the world is run by economists and accountants. If something affects the bottom line, it gets noticed. If it doesn't, it's fanciful. Biomimicry is affecting the bottom line in dramatic and positive ways. Most practically, using green chemistry and other biomimetic solutions eliminates liability, which is huge because the world is largely financed by insurance dollars, and insurance companies are always looking for ways to minimize exposure. Biomimetic approaches can also reduce energy use and pollution.

In an educated public, there's always an appetite to do the right thing. We're at that moment now, and the stakes are a lot higher than ever before. Even big corporations want to be seen as sustainable and are starting to support this kind of movement. And it's not just greenwashing; they genuinely want to be on the team. Biomimicry is fresh and exciting because there are new discoveries every day with a "Wow!" value that folks are noticing. People are tuning in; they want to know the next installment, the next best idea. If that next best product is both biomimetic and sustainable, that's a powerful driving force.

ABOVE *Venus fly trap.* OPPOSITE *The veiled stinkhorn* (Dictyophora duplicata).

🍄 MYCELIAL INTELLIGENCE 🍄

You can't discuss the innate intelligence of nature without acknowledging the mystery of mycelium and mushrooms. When you consider that our DNA is so similar to mycelium, and you start looking at the patterns by which mycelium grows, you discover that it's virtually the same as our nervous system, or our venous system, or even how the stars are disbursed through the universe.[4] Mycelium networks are not only equivalent in design to our own neural pathways but also similar in how they use electrolytes and electrical pulses.

The application of fungi's properties as medicine and for environmental remediation are advancing quickly, while a growing number of companies are using the properties of mycelium to create such industrial products as mats, building materials, and high-performance foams.[5]

> "A growing number of companies are using the properties of mycelium to create such industrial products as mats, building materials, and high-performance foams."

Does this suggest a single intelligence? To me, it's spooky in the most wonderful way when you see what mycelium do: They support life. They convert life. They carry life. They recycle life. They can heal you. They can clean up the environment. They can even shift our consciousness. And they can survive virtually anything. They are truly remarkable beings.

I recently read an article by some physicists who speculated about what species would take over the earth if humans became extinct. There were two schools of thoughts. One favored the cephalopods, like the squid and the octopus, because they're so intelligent. They've got the biggest eye-per-body-weight of any animal, they are easily trained, and their capabilities go way beyond those of any other creature on the planet. The other school of thought favored mycelium, not just because they're the most common species on Earth, but also because they are everywhere and are a part of every life-supporting function. There would be no plants on Earth, no animals, and no humans if there were no mycelium because we rely on mycelium for our very lives.

NOTES

1. Brendan B. Larsen, Elizabeth C. Miller, Matthew K. Rhodes, and John J. Wiens, "Inordinate Fondness Multiplied and Redistributed: The Number of Species on Earth and the New Pie of Life," *The Quarterly Review of Biology* 92, no. 3 (2017): 229-265, https://www.journals.uchicago.edu/doi/abs/10.1086/693564.
2. "Skin Cancer (Non-Melanoma): Statistics." Cancer.net, last modified January 2018, accessed July 2018, https://www.cancer.net/cancer-types/skin-cancer-non-melanoma/statistics.
3. "12th Annual EWG Sunscreen Guide." EWG.org, accessed July 2018, https://www.ewg.org/sunscreen/report/executive-summary/#;W1QReNJKhPY; "Are Sunscreens Safe?" *Scientific American*, accessed July 2018, https://www.scientificamerican.com/article/are-sunscreens-safe/.
4. Carl Zimmer, "Getting to Know Your Inner Mushroom," *National Geographic*, May 22, 2013, accessed July 2018, https://www.nationalgeographic.com/science/phenomena/2013/05/22/getting-to-know-your-inner-mushroom/.
5. The Mycelium Biofabrication Platform, accessed July 2017, https://ecovativedesign.com/.

TOP AND ABOVE *Mylo™*, a leather-like material made from mycelium cells, created by Bolt Threads.

OPPOSITE *Mycelium*.

🍄 FUNGI FACTS:
WEARABLE, BUILDABLE MYCELIUM

As researchers have become more familiar with the many properties of mycelium and certain mushrooms (not to mention their value as renewable resources), product applications have multiplied.

One of the first was hats and bags—natively popular in Romania—made from the soft, feltlike fibers of the amadou mushroom (known as *Fomes fomentarius* or tinder fungus for its historical use during the Ice Age to transport flame). In one of the latest innovations, German shoemaker nat-2 and the design firm Zvnder used amadou and mycelium's fibrous, malleable construction to produce "mushroom leather," which has been showing up in wallets and sneakers. Bolt Threads is turning mycelium into high-performance fabrics.

Given its strength, durability, and shape-shifting potential, an especially promising application has been the cultivation and processing of mycelium for industrial uses. Industrial designer Ecovative is testing products and prototypes that include panels, building blocks, furniture, and even foam. MycoWorks, founded by the installation artist Philip Ross, is beginning to create building materials out of agricultural waste using mycelium and certain mushrooms, such as reishi.

CHAPTER 3
THE WOOD WIDE WEB
☙ ❧ ☙

MERLIN SHELDRAKE

MERLIN SHELDRAKE graduated from Cambridge University with a PhD in tropical ecology and mycology. His research concerns the ecology of mycorrhizal fungal networks in Panamanian rainforests, where he worked as a Smithsonian research fellow.

While the internet is a useful metaphor for how fungal networks function in the earth, it doesn't quite capture their dynamic complexity. Each organic contributor is uniquely gifted and actively responds to changing conditions.

Fungi are fascinating because they can do things that other organisms can't do. They can digest things that other organisms can't digest. Some of them don't bother with sex. They're metabolically ingenious. Some of them can share genetic information with other fungi. They confuse our systems of taxonomy, making them hard to categorize but also fun to think about. Because of all this, there's a lot yet to be discovered about fungi, so there's a sense of excitement when exploring fungal networks and how they thread ecosystems together.

Mycorrhizal fungi are a broad category of fungus. They're called mycorrhizal because they relate to plant roots. *Myco* means "fungus" and *rhiza* or *rhizal* means "roots," so *mycorrhizal* means "root fungus." They grow very intimately into and around plant roots, and their mycelium extends outward into the soil.

Mycorrhizal fungi are very ancient, and nearly all plants have formed some kind of relationship with them.[1] They're thought to have been responsible for helping the ancestors of today's plantlife make it onto land about four hundred and fifty million years ago.[2] The ancestors of land plants were algae who lived in water. They were able to photosynthesize—to make energy from light and carbon dioxide—but they weren't so good at absorbing nutrients from solid substrates. When they started washing up on swampy shores, they encountered fungi that were able to digest solids, and the two began a relationship of exchange that persists today.

This is why mycorrhizal networks are often compared to the internet. Although that makes sense on the surface, fungal networks are much more complex and interesting. The most obvious distinction is that the internet connects people through hardware such as wires and routers and electromagnetic signals that, in themselves, are lifeless. Mycorrhizal networks consist primarily of "wetware"—the main actors aren't passive, but active. Unlike passive cables, they are agents with interests, making decisions moment by moment on how to manage their survival. The same goes for plants. The "Wood Wide Web" is a network of active players.

The phrase *Wood Wide Web* is a helpful one, because it communicates an idea of connected organisms very quickly. In Panama, I was studying a type of plants called mycoheterotrophs, which are unlike normal plants because they don't have chlorophyll, meaning they don't produce their own energy through the process of photosynthesis. The question then is how do they exist?

OPPOSITE *Coral fungus on a fir tree.*

They are somehow able to plug into fungal networks and acquire sugars and nutrients from other plants via the fungi.[3] Conventional green plants networked within the Wood Wide Web feed carbon-based compounds into the network and receive nutrients back from the fungi. It's a two-way exchange. These plants don't do that. They receive *both* nutrients and carbon from the network, which is why I joke that they're the hackers of the Wood Wide Web. Because all the carbon in the network ultimately has to come from plants, mycoheterotrophs show that carbon can pass between plants via the network.

Mycoheterotrophs are a great example of the many ways that different plants and fungi can behave within these networks. These are complex systems; there are lots of ways to engage. The relationship between mycorrhizal fungi and plants is generally thought of as mutualistic; that is, both partners benefit, although there are times when one plant or fungal partner might be getting more than the other. It's a flexible, fluid relationship that adapts to changing circumstances.

When we think about these exchanges, the question of what governs them comes up. How are these exchanges between plants and fungi regulated and negotiated? Humans tend to think of things in human terms. Some use "biological market" theories to explain fungi-plant interactions and think of plants and fungi as engaging in sanctions and investments and obtaining market rewards. Others grant more generosity to the players, describing them as caregivers who donate to and feed those who don't have enough—a more socialist framework. These concepts are somewhat useful but still human-sourced; they don't overcome the fact that we don't yet have a sophisticated understanding of how these fungal networks are behaving, which ultimately may not reflect a human framework at all.

These mysteries are yet another reason why mycology is fascinating—and not just to academics. I meet more and more people who are interested in the biological workings of the fungal kingdom simply out of curiosity or because they are hearing about some of the many potential applications of fungi. Many of my peers, for example, are concerned with issues of environmental degradation and sustainable futures. Some see mycology as part of the solution to a number of the problems we face. There are plenty of unanswered questions and the field is quite open, which makes the science and investigation a lot of fun.

NOTES

1. René Geurts and Vivianne G. A. A. Vleeshouwers, "Mycorrhizal Symbiosis: Ancient Signaling Mechanisms Co-opted," *Current Biology* 22, no. 23 (2012): R997-99, https://www.sciencedirect.com/science/article/pii/S0960982212012067.
2. Marc André Selosse, Christine Strullu Derrien, Francis M. Martin, Sophien Kamoun, and Paul Kenrick, "Plants, Fungi and Oomycetes: A 400 Million Year Affair That Shapes the Biosphere," *New Phytologist* (March 20, 2015), DOI/full/10.1111/nph.13371.
3. M. Sheldrake, N. P. Rosenstock, D. Revillini, P. A. Olsson, S. J. Wright, and B. L. Turner, "A Phosphorus Threshold for Mycoheterotrophic Plants in Tropical Forests." Proceedings of the Royal Society B: Biological Sciences (February 1, 2017), DOI:10.1098/rspb.2016.2093.

🍄 FUNGI FACTS:
BILLIONS AND BILLIONS OF SPORES

I am mesmerized by mushrooms' apparent immobility. When we look at them in the field, you see a fungal fruit body that seemingly is doing nothing. But if you look at a mushroom at the microscopic level and study what's happening underneath the cap, you see a flurry of spores being released from the gills on a continuous basis. We know from calculations that about thirty thousand spores are released from a single mushroom every second. That's billions of spores from a single fruit body every day of its often short existence!

A few years ago, some German scientists decided to estimate the total number of spores that mushrooms release into the atmosphere across the globe every year. Now, this wasn't guesswork. They followed chemical markers that the fungal spores carried on their surface into the atmosphere. These researchers calculated that tens of millions of tons of spores are ejected into the air every year, an amount approaching Avogadro's number (a unit of measurement that chemists use to count atoms and molecules): 10^{23} microscopic particles sent up into the air every year. I did my own calculations and determined that the combined surface area of all those spores is equivalent to the land area of the continent of Africa. It's not surprising that they can have a profound effect on our health and well-being.

Fungi are a huge and invisible part of our everyday environment, but we are a visual species. We're driven to appreciate and interact with organisms that are macroscopic, such as plants and animals, and yet we're surrounded by and immersed in fungi. Even when we understand the scientific facts about mushroom biology, there is so much about fungi we can't grasp. The mystery of their existence requires us to draw upon our imaginative powers to make sense of their enormous impact. You have to think very deeply to begin to appreciate the richness of our interactions with the fungal world. To really understand mycology, one almost needs to enter a spiritual realm.

—NIK MONEY

NIK MONEY is a professor of biology at Miami University in Oxford, Ohio, and author of numerous scientific papers and books about mycology.

ABOVE *Peziza sp.*

BELOW *Mushroom spores.*

OPPOSITE *Mycelium growing densely throughout a block of substrate.*

FOLLOWING PAGES *Unidentified mushroom species, Arctic National Wildlife Refuge, Alaska.*

CHAPTER 4
MUSHROOM MAD!

ஓ ஓ ஓ

GIULIANA FURCI

GIULIANA FURCI is an author, activist, and the founder and president of Fundación Fungi in Chile.

Fighting for the rights of fungi doesn't come to a person in a casual or even intentional way. It's a call to action that is felt, and there is nothing to do but answer it. Because it's not just about the health of fungi, but of all life.

Being completely mushroom mad is not something you choose. It's an overwhelming feeling of dependence on fungi, of devotion to fungi, of being at the service of fungi. I sometimes speak to people of *us*, of the fungi and our interconnection. It's not about being special or enlightened. For reasons I don't fully comprehend, I'm incapable of doing anything that doesn't in some way benefit our fungal kin.

After I gave birth to my son ten years ago, a friend called me in the hospital and asked, "What was it like?" And I said, "Do you remember what we felt when we found that puffball near Punta Arenas in Patagonia?" She said, "Yeah!" And I said, "It's like that! That's what it feels like." The only comparable feeling to giving birth for me is encountering certain species of fungi in the wild. It's just an overwhelming experience of plenitude and purpose. When women talk about Mother Earth and the throbbing of life and the forests, we speak with certainties that men don't have.

ABOVE *Purple* Laccaria amethystina *group.* OPPOSITE FROM TOP LEFT TO BOTTOM RIGHT *Giant puffball* (Calvatia gigantea); Pholiota *sp.;* Tapinella atrotomeutosa; Amanita muscaria.

THE LEGAL RIGHTS OF FUNGI

In Chile, we have what is called the National Environment Bill. It's a constitutional law at the highest legislative level that sets the rules for anything we do regarding the environment. Flora and fauna are mentioned, but the third, Fungi, from the third macroscopic kingdom, was missing. When you're looking at a plant, you are essentially looking at fungi; the two have never been separate. To speak of flora and fauna, is inaccurate when describing macroscopic life forms. They are three F's: flora, fauna, and fungi. If the third F were to be included wherever the first two are, the planet would be in a better place.

"No other organisms on Earth connect plants to animals, or bacteria to plants, like fungi do. Mycelium has the answers to saving our planet."

When that law was opened for revision, my colleagues and I at the Fungi Foundation, of which I'm the founder and president, saw an opportunity and decided to seize it. It took us two years to finalize the work that would translate to regulatory modifications for considering fungi a part of the environment, but we did it. In 2010, fungi were specifically added to Chile's general environmental regulations, with a mandatory requirement to inventory and classify them and to develop baseline fungal studies. By December 2013, it was a legal requirement for every environmental assessment to consider the impact on fungi before a development permit would be approved.

We're in the process of developing methodologies that will standardize how to perform and evaluate these impact studies. And because Chile is an environmentally diverse country with the driest desert in the world in the north, a largely Mediterranean climate in the center, and the tundra and glaciers of Patagonia in the south, we'll end up with three ecosystem templates that can then be replicated in other parts of the world.

The policy change we successfully advocated for has created dozens of workplaces for mycologists in our country. Right now, close to fifty consultants are working in the field on fungal baseline studies. When I started, there were none. We're also collaborating with different universities to develop certification programs for people such as botanists who are interested in this work, and we're working with the University of Concepción and the Ministry of the Environment to build on the twenty-species of native fungi already assessed using Red List of Threatened Species Criteria by the International Union for Conservation of Nature (IUCN).

Education is critical as this process unfolds, so we've produced two field guides (so far) for government officials and those who are working on these baseline studies. The first one was basic; the second one is more detailed with microscopic features of fungi and instructions for how to collect and study them. The guides help both those in the field and the governmental officials who will review the studies. They are also being used in universities by those who are going through certification processes. This second guide was reviewed by Paul Stamets as well as Jane Goodall. When Goodall discovered what we were doing back in 2013, she looked at me and said, "Don't stop. You're with fungi where I was with the chimps when I started." And I have taken that to heart.

🍄 EXPANDING THE REACH 🍄

Creating international partnerships is an important part of moving the fungi imperative forward. I'm an associate of Harvard's Farlow Herbarium, for example, and they've generously donated a library as well as a microscope and other resources to our foundation. The head of mycology there, Don Pfister, has taught workshops in Chile for the past four years. The Royal Botanic Gardens at Kew in England has also been working with us to describe new species and help put Chile on the map as a leader in the new discipline of conservation mycology. Other nongovernmental organizations (NGOs) around the world are now working to protect fungal networks including groups in Argentina and South Africa that are learning the strategies we used in Chile to integrate the science and value of this third kingdom into public policy.

And it is a strategy—a thought-out, top-down template approach that is deliberately replicable. But it requires maximum effort for maximum impact because little effort creates little impact, and with fungi that just won't work. Justice for the fungal kingdom needs a 100 percent effort if it is to be properly recognized, valued, looked after, and revered, as it should be.

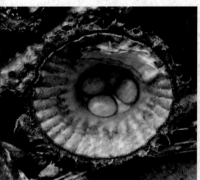  

🍄 SPORULATING THE FUTURE 🍄

Whether the work is being done at the grassroots community level or at the national and international policy levels, all routes lead to Rome. Each route is complementary and equally important because it's about the health of not just people but forests and the entire planet. It's full-spectrum health. That's what we've been doing in Chile: We are full-spectrum activists on environmental and education policies that create platforms for mycology to sporulate. We're not blossoming, we're sporulating!

If we have any hope of keeping planet Earth healthy or recovering areas that have been depleted by human activity, we must embrace an ecosystemic view of nature. And the only way to have an ecosystemic view of nature is by looking at the organisms that connect us all. Those organisms are fungi. No other organisms on Earth connect plants to animals, or bacteria to plants, like fungi do. Mycelium has the answers to saving our planet.

ABOVE LEFT *The fluted bird's nest fungus,* Cyathus striatus. ABOVE MIDDLE Tulostoma *sp.* ABOVE RIGHT Lactarius *sp.*

CHAPTER 5
CULTIVATING MYCO-LITERACY

PETER McCOY

PETER MCCOY is founder of Radical Mycology and author of *Radical Mycology: A Treatise on Seeing and Working with Fungi*. Radical Mycology is the theory and practice of using mushrooms and other fungi to create positive personal, societal, and ecological change. He is also the founder of Mycologos, the world's first school dedicated to mycology.

Mushrooms are a marvel. They serve us and other creatures in countless ways. But they remain a mystery. And as interest grows, so does the need for education.

One of the things I appreciate about teaching mushroom cultivation, mycoremediation, and ethnoecology is that wherever I go, I see all demographic groups coming together, drawn for many reasons. Whether it's the delicious food that mushrooms provide or the health, environmental, and community-building benefits in a world that's been damaged by human activity, fungi offer a pathway to a better future. We can create sustainable economies, manage local resources, reduce our carbon footprint, and clean and mitigate toxic soil with the help of fungi, but we need training and better data.

Fortunately, anybody can add to the science. The first step: Build your myco-literacy by understanding basic fungal biology and learning how to cultivate well through education and practice. As more and more people walk down that path, we'll propel mycology, mycoremediation, and all the other benefits of working with fungi forward with quantitative results that prove its value and give us something to say that people will listen to.

ABOVE *A mushroom identification workshop.* OPPOSITE Pholiota *sp.*

SPOTLIGHT ON: FUNGI ACTIVISM

The Radical Mycology Mycelial Network is an international, grassroots network of local groups and individuals that shares mycological information both with each other and within their respective communities. The following list presents just a few of the principles and practices that drive the network's mission to spread myco-literacy and activism as widely as possible.

INCREASE COMMUNITY RESILIENCE

- Create edible, medicinal, or guerilla mushroom installations at community gardens, food banks, and food justice organizations.
- Build relationships with and offer cultivation and educational support to local indigenous communities.
- For those working in labs, develop and maintain a library of fungal cultures or a spore bank to preserve genetics.

SUPPORT LOCAL ECOLOGIES

- Organize forest surveys to search for threatened mushroom and lichen species and use this information to support campaigns for forest protection.
- Increase species diversity and redundancy by spreading fungal spawn in any suitable area.

RECOMPOSE ORGANIC WASTE

- Develop relationships with local coffee shops. Use their spent grounds to grow mushrooms and give the mushrooms to local food banks or shelters.
- When clearing for a community garden, remove invasive plants and use this material as a mushroom substrate to build compost and topsoil.

EDUCATION AND OUTREACH

- Lead urban mushroom forays and note the abundance of wild food and medicine that can be harvested in urban areas.
- Hold awareness-building events to address local pollution issues or illegal polluting practices by industries.
- Teach workshops at schools: Children love to play with mushrooms and watch their mycelium grow.

GROUP-BUILDING ACTIVITIES

- Organize trips to mushroom farms to learn about the industry.
- Host mushroom-themed art parties and potlucks.

STRATEGY AND SUPPORT

- Build connections with local labs, public agencies, and universities to receive reduced-price or free testing of soil and water samples.
- Seek out experienced mycologists and social organizers to serve as mentors.

TOP LEFT *A hand holding a mushroom in substrate.* TOP MIDDLE *Boy inspecting mushroom.* TOP RIGHT *Gary Lincoff holding* Russula. OPPOSITE TOP *Girl picking a mushroom in the wild.* OPPOSITE BOTTOM LEFT *Mixed wild mushrooms.* OPPOSITE BOTTOM RIGHT *Paul Stamets holding up an edible* Russula.

CHAPTER 6
FUNGI: A BEE'S BEST FRIEND

STEVE SHEPPARD

STEVE SHEPPARD is the Thurber Professor of Apiculture and chair of the Department of Entomology at Washington State University, where he also heads the APIS Molecular Systematics Laboratory.

Bees are dying, industry fixes have failed, and food sources worldwide are threatened. Remarkably, one promising solution comes from an unlikely source: mushrooms.

I'm originally from Savannah, Georgia. My great-grandfather was a beekeeper who died when I was one, and as a small child I read his bee books and played with some of his equipment. As an undergraduate, I began taking some classes in beekeeping and then went to graduate school to work on pollination and honeybees. My PhD was on honeybee genetics and evolution. Bees have been a part of my life for as long as I can remember.

Honeybees are critical to the type of agriculture we have in this country. In the 1940s, there were about five million bee colonies in the United States; most farms were highly diverse, grew a variety of crops, and had a few beehives. There was also more unused land where other pollinators could breed and help out. Fast-forward to now, and we have closer to three million colonies and a lot more people. We are feeding bees through a different type of agriculture that depends on large monocultures of crops that need pollination. Once you have thousands of acres of a single crop, there just aren't enough native bees to pollinate that crop during the short, critical time it's needed. Honeybees are widely used to pick up the slack. Without them, we probably couldn't feed people in the way that we do now. Honeybees are by far the most important pollinators for agriculture. Approximately a third of all crops grown rely on cross-pollination to thrive.

Beekeeping in the United States reached a critical juncture in 1987; that's when the Varroa mite (also known as *Varroa destructor*) arrived, probably from Europe. It actually has a relatively harmless relationship with the Eastern honeybee (*Apis cerana*), but the Western honeybee (*Apis mellifera*) doesn't have good defenses against it. The Varroa mite feeds on the larvae, or baby honeybees, which spreads and amplifies the viruses it creates—in particular the deformed wing virus (DWV)—and reduces the longevity of the bees.[1] In largely temperate climates, this mite will keep multiplying until a colony dies, usually within two years. Colony loss from the Varroa mite (and other factors such as loss of foraging habitat, pesticides, and other viruses) averaged 40 percent across the United States from 2017 to 2018 and spiked as high as 85 percent in some states.[2] In Europe and Canada the losses have been more modest, but have reached 30 percent in some areas.[3]

OPPOSITE *Bee on a flower.*

"People don't really appreciate how dangerous our times are. We're at the cusp of an ecological collapse that could cause enormous poverty, starvation, wars, and political upheaval. If our ecosystems fail, what will people do for food? Our research on protecting bees will be fundamental to preserving biodiversity and saving our ecosystems, and my hope is that it will add a very important tool to our biological toolkit."

—PAUL STAMETS

The normal treatment used by large commercial beekeepers has been chemical control of the mite, but the mite has a very short lifespan and can develop resistance to these controls pretty quickly. Many products have come and gone because they are no longer effective. The only real sustainable solution for controlling these mites is through breeding more resilient honeybees and using agents that the mites are less likely to become resistant to. Another important consideration is that honeybees require a varied diet to be healthy. That's why the loss of functional habitat for honeybees is one of the underlying difficulties that the beekeepers and the honeybees face.

🍄 FUNGI TO THE RESCUE 🍄

A couple years ago, I became aware of the work being done by Fungi Perfecti and Paul Stamets. They had been testing some fungal extracts with antiviral properties, and so we began a series of experiments feeding these extracts to bees and had some success. The bees lived longer and their virus levels came down (a thousandfold or more in some instances), perhaps because the extract boosted their immune system, though we're not yet sure how it works. In a second set of experiments tested on five hundred hives in Washington State, we used an entomopathogenic fungus called *Metarhizium* that's been shown to harm the mite—it basically grows on the mites and kills them, and in certain dosages it doesn't affect the bees. *Metarhizium* itself has already been registered for treatment or control of other pest insects.[4]

Most recently we sent extracts of reishi mushrooms (*Ganoderma lucidum*), a staple in traditional Chinese medicine, and amadou (*Fomes fomentarius*), the "fire starter" mushroom that beekeepers use to smoke their hives, to the San Joaquin Valley in California, the largest producer of almonds in the world. Approximately eight hundred thousand almond trees bloom in a four-to-six-week period there in February. The only colony that can manage to pollinate those sorts of monocultures effectively and within such a short timeframe is a honeybee colony, which you can load on trucks and move to a particular location.

We inserted fungal extracts into 532 hives during the almond pollination season. I don't know if it was the largest bee experiment ever done, but it's the largest field experiment I know of—and it's the kind of real-world test that attracts attention, especially given that these hives are managed by commercial beekeepers. The experiment has generated a huge amount of data and samples for lab analysis that we're finally getting to the end of, and the results so far have been extraordinary. The antiviral effects of amadou against DWV have been on the order of 800:1; reishi has produced reductions of 45,000:1 against the Lake Sinai virus; chaga has reduced the Black Queen Cell virus at a level of 800:1. I have never seen such strong antiviral activity against bee viruses as I have seen with Paul's extracts. We submitted a paper summarizing our findings, and it was published in October 2018.[5] We hope the research will revolutionize our approach to protecting bees and help restore a healthy network of colonies worldwide.

Additional results from other studies have been equally promising. We have found, for example, that both the red-belted polypore mycelium and amadou mycelium significantly extended the life of honeybees. As an entomologist with more than forty years of experience studying bees, I am unaware of any reports of materials that can match this result.

ABOVE LEFT *Paul Stamets stands among hundreds of gallons of polypore mycelium extracts made at his lab complex at Fungi Perfecti, LLC in Kamilche Point, Washington.* ABOVE RIGHT *Polypore mushroom extracts are mixed with sugar water at a 1 percent concentration and poured into internal frame feeders.* OPPOSITE TOP *Beekeeping frame.* OPPOSITE BOTTOM LEFT *Bee on a raspberry flower.* OPPOSITE BOTTOM MIDDLE *Industrial farming.* OPPOSITE BOTTOM RIGHT *Dead bees.*

🍄 LIFE ON THE EDGE 🍄

By some measures, you might say we're on the edge. If the bees were to be lost, it would force radical changes in food costs and food security around the country. If the material that we're testing keeps working as it appears to be, the bees will be much better off, the beekeepers will be better off, and we'll all be better off. I've been working with industry and commercial beekeepers and hobbyists for many years, and they are desperately searching for answers. We think we may have some, and they have the potential to change everything.

NOTES

1. Myrsini E. Natsopoulou et al., "The Virulent, Emerging Genotype B of Deformed Wing Virus Is Closely Linked to Overwinter Honeybee Worker Loss," *Scientific Reports* 7 (2017): 5242; Kristof Benaets et al., "Covert Deformed Wing Virus Infections Have Long-Term Deleterious Effects on Honeybee Foraging and Survival," *Proc Biol Sci* (February 8, 2017), DOI: 10.1098/rspb.2016.2149.
2. "Preliminary results: 2017-2018 Total and Average Honey Bee Colony Losses by State and the District of Columbia," BeeInformed.org. accessed June 21, 2018.
3. "Reports of Bee Losses—U.S., Canada, and Europe," PRNHoneyBeeSurvey.com. September 6, 2015, http://pnwhoneybeesurvey.com/2015/09/reports-of-bee-losses-u-s-canada-europe/.
4. Donald W. Roberts and Raymond J. St. Leger, "*Metarhizium* spp., Cosmopolitan Insect-Pathogenic Fungi: Mycological Aspects," *Advances in Applied Microbiology* 54 (2004): 1–70, https://www.sciencedirect.com/topics/agricultural-and-biological-sciences/metarhizium.
5. Paul Stamets et al., "Extracts of Polypore Mushroom Mycelia Reduce Viruses in Honey Bees," *Nature: Scientific Reports*, 8, no. 13936 (2018). DOI: 10.1038/s41598-018-32194-8.

ABOVE *Bee on a flower.* **OPPOSITE FROM TOP LEFT TO BOTTOM RIGHT** *Gymnopus sp.; A tree frog peers over Mycenia in the Tambopata River region of Peru; Bee on a mushroom; Mushroom species unknown.*

CHAPTER 7

MYCOREMEDIATION: GROWING PAINS AND OPPORTUNITIES

DANIEL REYES

DANIEL REYES is a hydrogeologist and mycologist and the founder of Texas-based Myco Alliance.

The capacity of fungi to break down chemicals and eliminate toxins has been proven. The challenge now is finding innovative ways to put that ability into practice.

Mycoremediation essentially means the use of fungi to absorb, degrade, or sequester contaminants that have accumulated or been suddenly released into soil or water. They do this by breaking down complex hydrocarbons and chains of toxic molecules into small enough pieces that other microorganisms can finish the job, bringing life back into damaged sites.

There are basically two different levels of mycoremediation: low-tech, which represents the kind of approaches people would use at home, and high-tech, when you are dealing with a Superfund site or a large oil spill. In the case of something simple like motor oil that spills onto your driveway, you would first throw sawdust on the oil to soak it up, put that saturated sawdust in a tub, inoculate it with oyster mushroom mycelium, and then let the fungus do its work. If you try washing the oil off with a hose, it ends up in a stormwater drainage system that empties into the nearest creek and could end up in your drinking water.

Mycoremediation is not as simple as throwing mushrooms on a toxin, though—they need a wood- or grain-based substrate to grow in, which is why at home you would start with something like sawdust. This is one of the big lessons we learned from the Gulf of Mexico oil spill. You can't just throw mycelium into the ocean to absorb oil. Most mushroom mycelia are not salt-tolerant, and they need both oxygen and a substrate. There are marine-friendly species that we're beginning to explore, so there is hope, but the methodologies are still developing.

The other issue with mycoremediation is how to define restoration. Are you talking about returning a damaged site back to its original pristine condition, or settling for what may be legally acceptable but less than optimal for certain uses? For example, let's say you apply a mycoremediation strategy to contaminated soil and you now have a before and after treatment sample. You find that there's been a 96 percent reduction in polyaromatic hydrocarbons—a pretty good result. But is it good enough to grow an organic garden in? The next step would be to perform what is called an ecotoxicological test: adding something to the remediated soil such as plant seedlings or bean seeds or worms and then tracking what happens. From those results, you can determine the soil's true health. If some or all of the worms die in a few weeks or the seedlings deform or there's damage at a

OPPOSITE TOP Coprinopsis lagopus (© Taylor Lockwood). OPPOSITE MIDDLE Hericium coralloides (© Taylor Lockwood). OPPOSITE BOTTOM Marasmiellus sp.

cellular level, you may have cleaned the soil to a degree, but what will be able to live in it? What are the thresholds for what kind of uses? Here in Texas, the Texas Commission on Environmental Quality (TCEQ) sets standards for soil and water contaminants for different uses. If we ended up with that 96 percent result, for example, that soil would qualify for agricultural purposes under TCEQ standards, but I'm not sure I'd want to grow my own garden in it.

♣ SCALING UP ♣

After working for a number of years in the oil industry, I gained some insight into what pipeline companies look for when faced with a remediation and what environmental consulting companies generally offer to service those efforts. When I think about the role that mushrooms and mycelium could play at that scale, I come back to the need for pilot studies. A good place to create such experiments and develop new tools is at the city level.

Nearly every city in the country right now is dealing with the problem of stormwater runoff. Motor oil, roadside toxins, lawn and farming pesticides, and so on get swept into sewage systems without any filtration. One of the more promising approaches would utilize a series of mycelium-filled drip filters that are terraced below pipeline openings. The water coming out would go through a series of these boxes with the mycelium-infused substrate acting as a filter and pulling toxins out of the water, one after another, in a continuous filtration system down to the discharge point. You'd have to factor in a process for keeping the mycelium alive between storms, which would likely mean bringing experienced cultivators in periodically to manage and tend the filters, but that's why it's important to keep testing options. As such processes become standardized, they can be applied in other cities and seed other innovations.

🍄 COMMUNITY LABS 🍄

Another critical component to moving all this forward is the role of citizen science. On the one hand are grassroots folks who like to forage and cultivate; on the other are those who want to dig deeper into the properties of different mushrooms but aren't interested in academia or industry. They are often the ones who discover new compounds and applications, so the more tools and opportunities they have to study mushrooms, the better.

One idea that's starting to get noticed is creating community labs. They are based on the concept of a gym membership, where you pay a recurring fee for access to the gym and all its machines and exercise rooms. You can do cardio, you've got a treadmill, there are weights, maybe a pool or a basketball court. These mycology labs work on the same principle: Let's say I want to run some tests on contaminated soil and need a mass spectrometer to identify chemical markers. Maybe I need a flow hood to filter out bad air because there's a mushroom I want to study, but I don't have $2,000 to invest in one. Or I want to become a mushroom farmer, but I'm not interested in oysters or lion's mane because everybody is selling those. I want to develop some new strains. How great would it be if I had access to a climate-controlled growth space where I can develop and cultivate my new strain all the way through to fruiting? We're really on the cusp of what is possible.

ABOVE *Wild mushrooms being sorted for identification.* OPPOSITE *Xylaria hypoxylon (© Taylor Lockwood).*

NEXT-GEN EMPOWERMENT

It's important to emphasize that the science and application of mycoremediation is not being developed on an industrial scale as quickly as one might hope or expect, so it still depends on the contributions of grassroots groups and individuals. Over the last forty years, we've learned a lot from lab work and field tests showing that fungi can break down all kinds of chemicals and provide these incredible results, but progress on a larger scale is slow.[1] We still need to develop solid protocols and show good results to propel the science forward because as of 2018, there isn't a strong enough economic incentive for bigger players to engage, and so the funding isn't there.

I'm hopeful, though, that the new generation of mycologists will develop their own techniques and take the research we're doing and some of the methods we're coming up with to the next level. We're really just paving the way for them. They're watching our videos, they're reading our books, they're quickly catching up to what we're doing. I've taught a wide range of kids, from first graders to college students, and what amazes me each time is that they get it. These kids are really smart. They aren't slow to understand the processes we're describing.

NOTE

1. Shweta Kulshreshtha, Nupur Mathur, and Pradeep Bhatnagar, "Mushroom as a Product and Their Role in Mycoremediation," *AMB Express* 4 (2014): 29; Christopher J. Rhodes, "Mycoremediation (Bioremediation with Fungi)–Growing Mushrooms to Clean the Earth," *Chemical Speciation & Bioavailability* 26, no. 3 (2014): 196-98; T. Cajthaml, M. Bhatt, V. Šašek, and V. Mateju, "Bioremediation of PAH-Contaminated Soil by Composting: A Case Study," *Folia Microbiologica* 47, no. 6 (2002): 696-700.

🍄 SPOTLIGHT ON:
THE AMAZON MYCORENEWAL PROJECT

For decades, the Texaco oil company constructed hundreds of unlined earthen waste pits for its oil operations in Ecuador. Out of these pits, toxic waste has continuously leached into rivers and the surrounding environment. Damage to both the land and its people has been extensive. While various groups continue to fight the company in court over repatriation and the cost and responsibility of cleanup, the local people who've been affected by the spill created a nonprofit called the Amazon Mycorenewal Project (now CoRenewal) in 2010 that represents over 150 indigenous communities. With the help of a small team of consultants, they are researching, designing, and implementing biological solutions using the remedial power of fungi. Pilot projects are under way addressing both petroleum-contaminated soils and municipal wastewater contaminants, while the organization continues to study the capacity of native microbes to mitigate petroleum toxicity, with plans for a fungal-research laboratory and mushroom-cultivation space.

FUNGAL CLEANERS

Common fungi used in remediation and some of the contaminants they work on:
- Shaggy mane: arsenic, cadmium, and mercury
- Elm oyster: dioxins, wood preservatives
- Phoenix oyster: TNT, cadmium, mercury, copper
- Pearl oyster: PCB's, PAH's, cadmium, mercury, dioxins
- King oyster: toxins, Agent Orange
- Shiitake: PAH's, PCB's, PCP's
- Turkey tail: PAH's, TNT, organophosphates, mercury
- Button mushroom: cadmium
- King stropharia: E. coli and other biological contaminants

TOP *Taking notes about a foraged mushroom.*
OPPOSITE *Stropharia rugoso annulata.*

CHAPTER 8
THE MUSHROOM REVOLUTION IS HERE
❧ ❧ ❧

TRADD COTTER

TRADD COTTER is a microbiologist, mycologist, Environmental Protection Agency fellow, and cofounder (with his wife Olga) of Mushroom Mountain. He is the author of *Organic Mushroom Farming and Mycoremediation*.

There's a growing community of amateur mycologists around the globe who are unearthing the secrets of mushrooms and coming together to share what they know. And what we're finding out is changing lives.

What is more important than fungi? Here's my completely biased answer: Nothing. Fungi are the glue in the kingdoms of life that include insects, plants, and animals. They are what I call the first responders. They unlock doors. They are the keystone species. They make nutrients bioavailable for all the other organisms. They are hidden inside leaves, in our mulch, in the grass we walk on. They're on every continent and they're very resilient.

I was always curious about mushrooms, but I grew up in an area that is mostly myco-illiterate, where no one knew anything about them. So I decided to learn about them, one at a time. That evolved into starting mushroom clubs and leading forays. Then I started purchasing laboratory equipment and teaching people about mushrooms. Now I'm a mushroom researcher, a professional mycologist, a microbiologist, and one of the founders and owners of Mushroom Mountain, a mushroom farm and research facility.

When I teach mushroom cultivation, I tell everyone that, by default, they're going to be making soil. In upstate South Carolina where I live, for example, the topsoil was twelve to fifteen feet deep in the early 1900s. A study from a few years ago found that it's down to five to eight inches. It takes five to six hundred years to make an inch of this stuff, and we lost twelve feet of it in about one century. It's not rocket science that topsoil holds moisture. Civilizations come and go; those that can hold moisture are the ones that survive.

In the soil that is left—either in South Carolina or elsewhere—there are miles and miles of mycelium webbing. Airborne and waterborne contaminants such as particulates, parking lot runoff, and pesticides flow through that substrate, which acts as a beautiful micron filter that removes harmful bacteria within seconds. This is a big deal. The World Health Organization has reported that water-related diseases kill more than three million people a year, making them the planet's leading cause of death.[1] One in nine people lacks access to clean drinking water.[2] Imagine the possibilities if we could take portable mushroom kits to water-challenged places like Haiti and run their contaminated water through the mycelium, instantly trapping the bacteria and making the water safe. Well, that's exactly what we're doing.

OPPOSITE TOP *A young girl crossing a bridge in Haiti.* OPPOSITE MIDDLE LEFT *Cultivated maitake* (Grifola frondosa).
OPPOSITE MIDDLE *Reishi* (Ganoderma lucidum) *antlers.* OPPOSITE MIDDLE RIGHT *The garden giant* (Stropharia rugosoannulata).
OPPOSITE BOTTOM *Myconauts at Fungi Perfecti, LLC, preparing reishi kits for sale.*

Working with the Clemson University engineering department, we're in the process of converting small bins we'd been using in Haiti to grow biomass into water filters. The myceliated blocks of substrate in these bins essentially act as a micron filter. By creating entry points through these blocks, you can channel water through them and strip out the bacterial and biological contaminants—including cholera—that stick to the particulates that the mycelium filters out. Haiti won't be replacing conventional water-treatment plants anytime soon, but as a low-tech solution to a huge problem, it's really a potential breakthrough.

> *"I'm seeing a revolution, a wave of mycelium activism splashing into the world. We're seeing more clubs, more mushroom hunters, and more access to information than ever before."*

🍄 IT'S THE MUSHROOM, MON 🍄

The more that I study mushrooms and fungi, the more I feel spontaneously obligated to travel abroad and help those in need. Back in 2015, I was invited to Jamaica to teach workshops on growing mushrooms. Jamaica imports $35 million of mushrooms every year, and there wasn't a single reliable grower or spawn lab on the island. Not surprisingly, every workshop sold out. We went to Jamaica a couple more times, and in 2017 we met with the prime minister and minister of agriculture. The result is that there is now a mushroom laboratory and tropical mushroom research station in eastern Jamaica. This is so important because not only are Jamaicans starting to grow their own mushrooms, but sugarcane termites are a major problem on the island. Fifty percent of all the cane on the east side of the island is destroyed by termites every year. Sugarcane is hand-harvested, and the fields are sprayed with pesticides while the workers are in them. It was a terrible situation.

After we taught the locals how to search for and analyze mushrooms using the new research station, they found a termite-killing fungus within six months. Great news for them, bad news for the pesticide industry. And that's why we created the lab, so we can look through the woods, find specimens, and figure out what to do with them. Jamaica is uncharted. There are so many different fungi on the island that haven't been recorded. What are they capable of doing there? What can we do with them here? How many other countless places are similarly rich in possibility?

If you're someone who knows how to create food out of nothing, how to filter water, how to create medicinal compounds, and you take that knowledge to a place like Jamaica, they look at you like you have superpowers. When they look like that at me, I tell them, "You're about to have superpowers, too. And when you learn those powers, it's your obligation to teach the next person, the next village. That's how it works. Protect and share the mycelium that you find. It's precious."

OPPOSITE *Paul Stamets and Michael Vallo examine a small* Polyporus elegans.
ABOVE *Specimens of* Cortinarius traganus *group are cut in half to examine interior flesh.*

A COMMUNITY OF CHANGE AGENTS

When I started in this field twenty-five years ago, I knew nothing about fungi—but it turned out to be a great teacher. It taught me observation. It taught me to slow down. It taught me to put my cell phone down and pay attention to small things. Studying fungi has been humbling and grounding, reminding me that I'm a member of the planet's dominant species, which keeps growing and taking and not giving back.

Since I've been teaching, I've seen a lot of new talent—people thinking way outside the box with access to new and inexpensive technology. I've met hydrogeologists making mushroom filters and a person studying how to break down cigarette butts and oil spills using fungi. I have an intern who's working on a fungus she found in Guatemala that breaks down acrylic and cellophane. It's amazing.

I'm seeing a revolution, a wave of mycelium activism splashing into the world. We're seeing more clubs, more mushroom hunters, and more access to information than ever before. We need this new generation to sustain and expand what we know and to create beneficial new skill sets, both in the lab and in the field. That's what happened to me. I followed some great mycologists and their work and they inspired me, and they were always willing to share their information because that's how they started too.

The first mushroom book I read was *The Mushroom Cultivator* by Paul Stamets. My first mushroom-cultivation manual was a three-page typewritten letter I saw in the back of a *High Times* magazine. Information just wasn't around. Today, with the internet, it's everywhere, *and* it's free. It's not always accurate, but plenty of people are modifying that knowledge base on a nearly daily basis. The more books and blogs and websites we have, the more people who are growing, the more people who are building on that knowledge, the better. There are an estimated five million species of fungi on the planet, but we know so little about them.

There's this great book from the 1960s called *Mushrooms, Molds, and Miracles* by Lucy Kavaler. In the first few pages, she writes, "If all the mushroom people can work together and share their information, there can be a huge transformation, so much transformation that there will be no starvation and no war on the planet. So, what are we waiting for?"[3]

Part of my mission is that call to action. I wouldn't call myself a militant mycologist, but it's close. It's time to start acting, to fix this planet of ours.

NOTES

1. Jessica Berman, "WHO: Waterborne Disease Is World's Leading Killer," VOA News, October 29, 2009, accessed July 2018, https://www.voanews.com/a/a-13-2005-03-17-voa34-67381152/274768.html.
2. "Progress on Sanitation and Drinking Water: 2015 Update and MDG Assessment," UNICEF and World Health Organization, accessed July 2018, http://files.unicef.org/publications/files/Progress_on_Sanitation_and_Drinking_Water_2015_Update_.pdf.
3. Lucy Kavaler, *Mushrooms, Mold, and Miracles: The Strange Realm of Fungi* (New York, NY: The John Day Company, 1965).

CHAPTER 9

THE SKY IS FALLING, BUT THERE'S A NET

ରେ ରେ ରେ

PAUL STAMETS

PAUL STAMETS is the preeminent mycologist in the United States. He has discovered several new species of mushrooms, pioneered countless new techniques, published several best-selling books, and won numerous awards.

The story of fungi is the story of our essential unity. The more we understand and respect this third kingdom, the more solutions we'll uncover for the many challenges that face us. It's not surprising that we are latecomers to the discovery of the usefulness of mushrooms, because our encounters with them are very short; they pop up suddenly and then they're gone. With plants and animals, by comparison, which are visible for weeks, months, years, the embodied knowledge we have of them is naturally much greater. Mushrooms can feed you and heal you, they can kill you, and they can send you on a spiritual journey. Then they disappear. It's very hard to understand them, but understand them we must.

It doesn't really help to tell people that the sky is falling or the soil is collapsing without some options for solving the situation. This is what fungi and mycelium bring to the table: They represent ecological remedies that literally lie beneath our every footstep. To take just one example, consider agriculture. The new standard for sustainability is "no-till" farming, a process that features minimal disturbance of the soil in order to maintain its ability to absorb water and sequester carbon. How does one do that? By keeping the mycelium intact and planting seeds that are dusted with mycorrhizal fungi. Ten years ago, you couldn't find a single bag of such seeds; now you can't find any without it. Ninety percent of the soil now sold in garden centers are fortified with mycorrhizal fungi.

The qualities of these fungi have largely gone unnoticed in the past, but we are noticing them now. We've identified several hundred species and cultures that have different talents for solving a lot of environmental challenges, and that's just the beginning. It's a matter of matching those talents specifically with a targeted concern. And as we keep doing that, we are building what I call an armamentarium of solutions, which is unprecedented in our time. We can dial in solutions using fungi with specific skillsets to address many of the environmental challenges we face.

This has been my mission all along, and it turns out that a lot of my early ideas have proven to be factually correct and scientifically supported. When I do research experiments, I look for my biggest critics, often those working in universities, and invite them to prove me wrong. "Do this research on your own," I suggest. "Design the experiment and the protocol, test the hypothesis, and get back to me." Many of my most vocal skeptics have become big supporters.

OPPOSITE TOP LEFT TO BOTTOM RIGHT Auriscalpium vulgare; Dictyohpora mulitcolor; *Mushroom species unknown*; Thaxterogaster porphyreum; Podoserpula pusio; Cordyceps militaris; Calostoma insignis; Austroboletus betula; Mycena spinosissima. *All images on page © Taylor Lockwood.*

And so we shouldn't underestimate the intelligence of nature and the many organisms that have evolved to this day. We are here today because of very smart choices made along the evolutionary path and because of our ability to communicate with other species and enlist them to our benefit. And this doesn't just work for humans. The communities of nonhuman species working together across the planet are the true architects of the ecosystems that maintain the circle of life.

🍄 STRENGTH IN NUMBERS 🍄

What's been missing in the scientific conversation is the impact of the third kingdom, the mycelial and fungal networks. They are the foundation of the food web. Seventy percent of the biological carbon in soil is fungi, both living and dead. I've been attending conferences for years trying to engage the scientific community in acknowledging this vital piece of the puzzle, and it took a long time to register. Everyone's looking aboveground for solutions to our environmental problems. They're looking at oil, at concrete, using mechanistic and conventional ways of thinking without understanding that the biology of nature has already solved many of these problems. Unless we come to terms with our fungal underlords, our biological myopia may be the downfall of us all.

Humans indulge the idea that we are the highest species, the top of the food chain, and that the main purpose of the biosphere is to support us. But is this "highest species" worth protecting? This illusion of biological grandeur is the cause of our suffering because of the egocentricity it covers up. We're still very much in kindergarten when it comes to understanding how to cocreate a sustainable future for all beings—and all beings are necessary to make that future possible.

A core concept of evolution is that, through natural selection, the strongest and fittest survive. In truth (and scientifically proven), communities survive better than individuals, especially communities that rely on cooperation. Acting on such a principle, people want to give in order to receive, which I think reflects the power of an essential goodness. I see this as a major force of nature; that evolution is based on the concept of mutual benefit and generosity. It's certainly true in the world of fungi and plants and trees. Yes, there are negative influences in the human domain. We do have some bad actors. But I think their appearance reminds the majority that it's far better to base our actions on the values we strive to attain, on the principles we want to carry forward into future generations. Who wants to teach their children to be greedy, mean, and violent? Benevolence, kindness, trust, forgiveness—the fact that these concepts exist is to me de facto proof that the evolution of life has been based on the concept of goodness.

Fungi builds soil, which expands the carrying capacity of the ecosystem and thus the capacity for adding biodiversity. The more complex your biodiversity, the larger your cast of characters, the more opportunities for plants and other creatures to collaborate on providing maximum benefit to all. It's like anything else: The larger your selection pool, the more likely your successes. True wealth is not measured in material possessions but by the abundance of options and choices.

Mycology (the study of fungi) is capable of offering so many solutions. Yet the field is underfunded, underappreciated, and underutilized. I think this shows the chasm of knowledge we have to overcome. But it's also a perfect opportunity to create an "integrative science," building bridges of

OPPOSITE *One healthy specimen of* Morchella conica *or black morel mushroom.*

knowledge between researchers of fungal networks and other scientists to develop an applied mycology grounded in practical solutions. We need to be engaging fungi purposefully, because they are the agents of sustainability, they will give us resilience.

🍄 THE TRUTH OF OUR INTERCONNECTEDNESS 🍄

"With fungi and mushrooms, we've only scratched the surface of how they work and what they know."

The problem we face today is that we live in artificial realities wherein we invent our own facts. The more artificial the reality (for example, "Climate change is a hoax"), the more divorced from the truth one gets. These bubbles of false realities are often constructed to serve the interests of certain groups and people who are trying to hijack your consciousness. It's like getting stuck in a video game. You end up believing in and playing by the rules of that game, but the problem is that those rules don't apply outside the video game. Once you embed yourself in nature and start working in the real world, you'll encounter a new set of well-established rules and a knowledge base that's been tested for millennia. We won't solve our most pressing problems unless enough people leave the video games they're immersed in and fully reengage in the consciousness of nature.

We all have a deeply embedded yearning to experience nature. Can you imagine what someone would be like if they had never been exposed to the natural world? If their reality had been limited to bricks and buildings, to straight lines and asphalt? If that is your only view of reality, how many truly creative ideas can come from that? Unfortunately, most people are, to some degree, ecologically impoverished, aptly coined the "nature-deficit disorder" by Richard Louv. But it's not too late. We can still use our best technology and the tools of science to respectfully and humbly unlock nature's secrets and reveal the depths of her knowledge. With fungi and mushrooms, we've only scratched the surface of how they work and what they know. There are literally thousands of unknown species, each with unique properties that may hold answers to the massive challenges we face.

Five or ten years ago, the idea that "everything is connected" sounded like New Age nonsense. Now it is less controversial, but we still need to develop more ways to show that this interconnectedness is real and that anyone can participate in a solution and have a meaningful impact.

🍄 A TIME FOR THERAPEUTIC INTERVENTION 🍄

Like it or not, we're actively involved in our evolution. The challenge is that we have to evolve faster because the problems we're creating are eclipsing our ability to adapt. And if we don't adapt and get ahead of this curve, we will lose the race and our species will instead simply race toward extinction. It's happened to countless species before ours. I do think it's a challenge we're well equipped to meet, but it will require a different mindset, a paradigm shift in consciousness, that looks at everything holistically.

When you consider the webs of dark matter showing up in hi-tech scans of the universe, the webs of mycelium branching out through the planetary biosphere, the webs of capillaries in our bodies, and the neural structures in our brains—these to me are central to this new paradigm. It's all connected, it's all coevolving. Networks are the rule of nature; not the exception.

OPPOSITE TOP *The rare and edible lion's mane mushroom.* OPPOSITE BOTTOM *Ramaria sp.* FOLLOWING PAGES *Gymnopus erythropus.*

SECTION II
FOR THE BODY

We do more than make mushrooms . . .
We have the ability to do so much more than
just break down matter . . .
Like the fruit of our labor, most of you have only
scratched the surface of our usefulness.
We are the changers . . .

Mushrooms have long been savored for their culinary virtues, but few people are aware of their nutritional and medicinal value. From Egyptian nobles and Greek physicians to Toltec shamans and Buddhist monks, mushrooms have been used for millennia by cultures around the world for healing, ceremony, and survival.

Led by a growing community of citizen mycophiles, activists, and innovators, the study and cultivation of mushrooms is slowly uncovering their incredible complexity and capacity to treat an astonishing range of diseases and conditions. The beauty is that we've only begun to scratch the surface of their potential.

In this section, experts explore the many potential health benefits and delicious nature of mushrooms.

MUSHROOMS: PHARMACOLOGICAL WONDERS • ANDREW WEIL *(pg.74)*

SPOTLIGHT ON: A "TAIL" OF RECOVERY • PAUL STAMETS
FUNGI FACTS: THE HEALTH BENEFITS AND VARIETIES OF MEDICINAL MUSHROOMS

FUNGI AS FOOD AND MEDICINE FOR PLANTS (AND US) • EUGENIA BONE *(pg.80)*

FUNGI FACTS: THE FIRST CULTIVATED MUSHROOM

A FORAGER'S HOMAGE • GARY LINCOFF *(pg.92)*

SPOTLIGHT ON: GARY LINCOFF: MYCO-VISIONARY (1942–2018)

A MUSHROOMING OF INTEREST • BRITT BUNYARD *(pg.96)*

SPOTLIGHT ON: DESERT TRUFFLES • ELINOAR SHAVIT
FUNGI FACTS: MUSHROOMS ACROSS TIME: AN ANCIENT HISTORY • ELINOAR SHAVIT

GROWING YOUR OWN • KRIS HOLSTROM *(pg.104)*

FUNGI FACTS: EATING WILD • KATRINA BLAIR

MUSHROOMS TO THE PEOPLE • WILLIAM PADILLA-BROWN *(pg.108)*

OUR GLOBAL IMMUNE SYSTEM • PAUL STAMETS *(pg.112)*

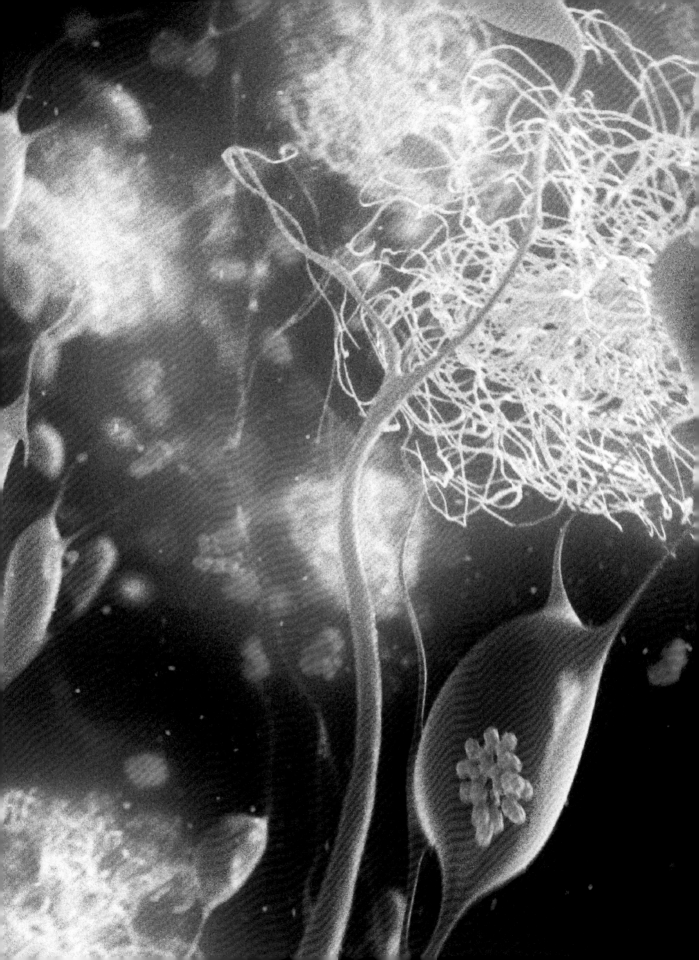

CHAPTER 10

MUSHROOMS: PHARMACOLOGICAL WONDERS

ANDREW WEIL

ANDREW WEIL, MD, is a world-renowned leader and pioneer in the field of integrative medicine, a healing-oriented approach to health care that encompasses body, mind, and spirit.

Western physicians aren't trained to consider mushrooms as having legitimate medicinal value; but they do, and in some cases they are more effective and appropriate than traditional approaches. Fortunately the tide is turning, and their benefits are beginning to find converts.

Here's a question: Why do mushrooms and psychoactive plants, such as opium poppies and cannabis, produce molecules that fit receptors in the human brain and body? What does that say? What does that mean? Is this coevolution? Have these receptors developed in us because of our relationship with these organisms, because we're supposed to be using them in some way? When you try to ask questions like this in medical school, the lecturer always says, "That's teleological," meaning the question is about final causes and purposes and therefore we don't have to answer them. But I've always thought that we need to try.

Mushrooms are not plants. They are in their own kingdom. In fact, they're more closely related to us than they are to plants. We share more DNA sequences with mushrooms than we do with plants. Both humans and mushrooms come from the same branch of the evolutionary tree, separate from the branch that led to plants. Mushrooms are completely unusual organisms. Their tissue is more like animal tissue than plant tissue, and their nutrient content is also more like animals than plants. Could some of the medicinal effects they have on us be explained by this close genetic relationship?

My own experiences with mushrooms over the years have been very beneficial and a great advantage to me as a physician, especially in making me more aware of the natural world. I came into medical school with a degree in botany, which was very unusual, but it inspired me to make more use of natural remedies and natural approaches in healing, prevention, and treatment. My experience with mushrooms on all levels, whether it's been learning to identify them, foraging for them, tasting them, cooking with them, eating them, or experimenting with their medicinal and consciousness-altering effects, has been a very important part of my development, both personally and professionally.

They call to me. They talk to me. I enjoy them. I like living around them. I like taking care of them. I enjoy their beauty. I like making them a major part of my diet. I especially like using them as remedies, because mushrooms (as well as plants) have complex chemistries that are quite different from the single extracted molecules purified by pharmaceutical companies into various supplements and drugs. Giving the body these complex arrays of molecules as nature produces them interfaces with the machinery of our bodies in a better way. I've seen the quality and effects of these perform quite differently from and often better than isolated chemical drugs. Chemical drugs have their place, but it's often more advantageous to use a complex natural remedy.

OPPOSITE *Rendering of neural connections.*

✿ HEALTH BENEFITS ✿

In my talks, I've sometimes remarked how odd it is that conventional scientific and medical communities have ignored mushrooms as possible sources of new medicines. Perhaps it's because of their unusual chemistries. They have molecules not found elsewhere in nature. Some are filled with toxins. But if you're looking for pharmacologically active agents, you go where there are toxins because many of those toxins become useful medicines in low doses.

There's a long tradition of using mushrooms as medicines in East Asia, especially in Korea, China, and Japan. I've studied traditional Chinese medicine and was very impressed with its use of mushrooms to fill niches for which we don't have solutions in Western pharmacology, like enhancing immune function or helping the body defend against cancer in ways that are not harmful. We don't have things like this in Western medicine.

I teach other doctors about the uses of mushrooms, and I recommend mushrooms and mushroom products frequently to patients; their health benefits are myriad, and we keep finding new ones. The possibility that they can augment conventional cancer treatment is especially intriguing.

Turkey tail (*Trametes versicolor*) is a good example. It's an unassuming mushroom—small and thin with gray and brownish ribbon coloring—that grows on dead logs and trunks in the forest. It's also one of the best-researched and most impressive of the immune system-enhancing mushrooms, having been used in China and among Native Americans for hundreds of years.[1] I take it myself. It's completely nontoxic, and I recommend it to many people, especially as an adjunctive treatment for anyone undergoing conventional treatment for various cancers.[2] For people who are susceptible to infection, who have cancer, or who are undergoing chemotherapy or radiation that can damage bone marrow and the immune system, turkey tail can be very protective.

Some mushrooms may be useful for people with degenerative nervous system diseases or nerve injuries. Lion's mane is one of them. It's an edible mushroom that you can find in forests, and it's delicious. Japanese growers learned how to cultivate it, and it's now available in many markets there. It also has a nerve growth factor in it that's unique. I often recommend lion's mane to people who have nerve injuries, nerve compression injuries, or neuropathy. There is also a possibility that it has a preventative and therapeutic effect in people with dementia.[3] And lion's mane is not alone. A 2017 study of eleven mushrooms, including maitake, cordyceps, and nine other edible and medicinal species, found that each one increased the production of specific nerve cells in the brain that could protect against both dementia and Alzheimer's.[4] There is also evidence that the high concentration of antioxidants in mushrooms minimizes cell damage, slows down the aging process, and lowers the incidence of neurogenerative diseases.[5]

Most of the mushrooms that I've been referring to are called polypores—shelf fungi with pores or gills on the underside that grow on living or dead trees. As a group, they have few poisonous members (the species *Hapalopilus nidulans* is one) and many are documented to be quite safe to experiment with. They represent an impressively wide range of interesting medicinal possibilities, and since we don't really have anything for some of these conditions, we need to keep testing their compounds.

I have a good friend and colleague who's a physician trained in Chinese medicine. I've traveled to China with him and brought him to the University of Arizona, where I'm based, to teach our

OPPOSITE *A small blue mushroom, species unknown.*

doctors. I once heard him say, "If I could summarize the purpose of all Chinese medicine in one sentence, it would be to dispel evil and support the good." In Western medicine, we're entirely concerned with dispelling evil. We spend all of our effort on identifying and eliminating agents of disease. That has its place, but we do almost nothing about supporting the good, which is supporting the body's healing and defensive functions. We need to balance both of those approaches, and a lot of work with mushrooms is in the area of supporting the good through discovering mushrooms' ability to enhance our immune and regenerative systems.

🍄 A GROWING ACCEPTANCE 🍄

In the English-speaking world, mushrooms have been regarded quite negatively. They're seen as having no nutritional benefits or medical potential, and we've been very afraid of their psychoactive properties. I think all that's changing. There's been a reevaluation and growing acceptance of mushrooms in Western culture. More people are including edible mushrooms in their diets, and I think we'll see psilocybin mushrooms become less stigmatized. I also think the medical profession is slowly opening up to the great value that mushrooms offer for health and healing.

NOTES

1. A. Weil, "Turkey Tail Mushrooms for Cancer Treatment?" Drweil.com, April 1, 2011, accessed July 2018; F. Li, H. Wen, Y. Zhang et al., "Purification and Characterization of a Novel Immunomodulatory Protein from the Medicinal Mushroom *Trametes versicolor*," *Science China Life Sciences*, vol. 54 (2011), https://doi.org/10.1007/s11427-011-4153-2.
2. C. J. Torkelson, E. Sweet, M. R. Martzen, M. Sasagawa et al., "Phase 1 Clinical Trial of *Trametes versicolor* in Women with Breast Cancer," *ISRN Oncology*, vol. 12 (2012), http://doi.org/10.5402/2012/251632; Q. Zhang, N. Huang et al., "The H+/K+-ATPase Inhibitory Activities of Trametenolic Acid B from *Trametes lactinea* (Berk.) Pat, and Its Effects on Gastric Cancer Cells," *Fitoterapia* 89 (September 2013): 210-17, https://www.sciencedirect.com/science/article/pii/S0367326X1300141X?via%3Dihub.
3. Erica Julson, "9 Health Benefits of Lion's Mane Mushroom (and Side Effects)," Healthline.com, May 19, 2018, accessed July 2018, https://www.healthline.com/nutrition/lions-mane-mushroom.
4. Chia Wei Pfan et al., "Edible and Medicinal Mushrooms: Emerging Brain Food for the Mitigation of Neurodegenerative Diseases," *Journal of Medicinal Food* (January 2017).
5. Matt Swayne, "Mushrooms Are Full of Antioxidants That May Have Anti-Aging Potential," *Penn State News*, September 9, 2018, https://news.psu.edu/story/491477/2017/11/09/research/mushrooms-are-full-antioxidants-may-have-antiaging-potential.

SPOTLIGHT ON:
A "TAIL" OF RECOVERY

ABOVE *Paul Stamets and his mother.*

My eighty-three-year-old mom was diagnosed with advanced Stage 4 breast cancer in 2009. I spent a lot of time in oncology clinics with her and talking to doctors who gave her less than three months to live. She had tumors throughout her body and was too old for a mastectomy or radiation therapy. One of the docs mentioned an interesting study on turkey tail mushrooms at Bastyr University with the University of Minnesota Medical School and that mom might want to try taking those. She laughed and said, "Oh, my son is supplying those!" So she was put on Taxol (from the Pacific yew tree) and Herceptin (a conventional antibody drug), and she started taking eight turkey tail capsules a day: four in the morning and four in the evening. In 2018, nearly a decade later, my ninety-three-year-old mother was alive and well with no detectable tumors. It has been widely reported in the literature that medicinal mushrooms like turkey tail help conventional medicine and chemotherapeutic agents work better, and I think she's a living example of that. I don't know the influence each of those substances had, but I'm certainly grateful.

—PAUL STAMETS

 **FUNGI FACTS:**

THE HEALTH BENEFITS AND VARIETIES OF MEDICINAL MUSHROOMS

The health benefits of mushrooms go back thousands of years. The birch polypore *Piptoporus betulinus* (*Fomitopsis betulina*), for example, known for its antibiotic, antiparasitic, and anti-inflammatory properties (among other benefits), was found in a Neolithic medicine kit that Otzi, the Ice Man, carried 5,300 years ago.[1]

In the modern age, the research-based evidence for the health benefits of various mushrooms continues to grow, showing their ability to:

- Empower the immune system
- Help defend against diseases and infection caused by viruses, bacteria, and protozoa
- Stimulate cell-damaging free radicals
- Increase bone strength and durability
- Stimulate tumors to self-destruct
- Support nervous system, neurogenesis, and mental health
- Augment longevity
- Modulate blood pressure and cholesterol
- Help normalize blood sugar levels
- Support overall health by providing a rich source of vitamins, minerals, antioxidants, amino acids, fiber, and protein

Many of the mushrooms with the most powerful healing agents are not native to the United States; Asian mushroom varieties are especially high on the healing list. And not all medicinal mushrooms are edible.

These are the varieties that have shown the strongest health impacts—so far:

- Agarikon (*Fomitopsis officinalis*)
- Almond (*Agaricus subrufescens*)
- Black hoof (*Phellinus linteus*)
- Chaga (*Inonotus obliquus*)
- Chanterelles (*Cantharellus cibarius*)
- Cordyceps (*Cordyceps militaris*)
- Cremini (*Agaricus bisporus*)
- Lion's mane (*Hericium erinaceus*)
- Maitake (*Grifola frondosa*)
- Oyster (*Pleurotus ostreatus, Pleurotus pulmonarius*)
- Reishi (*Ganoderma lucidum, G. linzhi, G. resinaceum*)
- Shiitake (*Lentinula edodes*)
- Turkey tail (*Trametes versicolor*)
- White beech / Shimeji (*Hypsizygus tessellatus*)

NOTE

1. M. Pleszczynska et al., "*Fomitopsis betulina* (Formerly *Piptoporus betulinus*): The Iceman's Polypore Fungus with Modern Biotechnological Potential," *World J Microbiol Biotechnol* 33, no. 5 (May 2017): 83, https://www.ncbi.nlm.nih.gov/pubmed/28378220.

CHAPTER 11

FUNGI AS FOOD AND MEDICINE FOR PLANTS (AND US)

EUGENIA BONE

EUGENIA BONE is a nationally known nature and food writer. Her work has appeared in many magazines and newspapers, including the *New York Times*, the *Wall Street Journal*, *Saveur*, *Food & Wine*, and *Sunset*. She is the author of six books, including *Mycophilia: Revelations from the Weird World of Mushrooms* and her latest, *Microbia: A Journey into the Unseen World Around You*.

The complexity of the unseen world continues to amaze me, and I have mushrooms to thank for that. They helped me understand a bit of the complex and beautiful workings of nature, and opened my eyes to the vast interconnectedness of all living things.

I got into mushrooms because I love to eat them, especially wild mushrooms. But if I was going to find the right ones, I had to learn something about their biology and how and where they grow. There's a huge subculture of mycophiles, people who are fascinated by mushrooms. They party together, they travel together, they hunt and eat mushrooms together. They are sort of bloated pleasure-seekers with a scientific bent. My kind of crowd. So I started hanging around these folks and attending their amateur mycology festivals and forays in order to become a better mushroom hunter. In the process, I fell in love with mycology, the biology of fungi. And that changed the way I saw everything. Mushrooms are the window through which I came to understand nature in a deeper way. It was the story of mushrooms, of fungi, that introduced me to the glorious story of symbiosis, the interdependence of all organisms, both seen and unseen.

ABOVE *Chopping mushrooms for a meal.* OPPOSITE FROM TOP LEFT TO BOTTOM RIGHT *White shimeji* (Hypsizygus tessulatus)*; Prepared mushroom dish; A variety of chopped mushrooms; A basket of one of the most delicious of all mushrooms, porcini* (Boletus edulis *group*)*.*

A MYSTERY UNFOLDING

Like many folks, when I first started learning about mushrooms I didn't realize they weren't plants. The mushroom is neither vegetable nor animal but somewhere in-between, a separate kingdom altogether. They are mostly water and fiber like plants, and are reproductive organs like fruits, but evolutionarily speaking, they are closer to us on the tree of life. There are as many as 3.8 million species of fungi—many times more than plants.[1] About twenty thousand of them produce mushrooms, and of those, a small number are seriously poisonous, a slightly larger number (maybe thirty or so) are known to be edible, and a handful of others have powerful health benefits. The rest are mostly unknown.

Maybe it's that mysteriousness that makes so many people afraid of mushrooms. They associate mushrooms and fungi with mold, death, and decay—which is understandable. Many types are rotters, after all. And certainly, there are some that, if you swallow them, can take out your liver or your kidneys. But there are berries in the woods that can kill you as well. So it's really just a matter of knowing your mushrooms.

Mushrooms are the fruiting bodies of fungi. Imagine it like this: The fungus is like an underground apple tree, and the mushroom is like the apple. They are too small to have stomachs, so they eat by excreting enzymes—kind of like your gastric juices—that break down food outside their bodies. A fungus living in a wood chip will break the plant matter down to its molecular parts—carbon, nitrogen, phosphorus, and so on—and then the fungus will absorb the nutrients it needs, which allows it to grow a little bigger. There seem to be fungi that can break down anything that's hydrocarbon based, and those that decompose dead and dying organisms actually move those nutrients back into the ecosystem. Fungi act at the end of the lifecycle of the organisms they break down, but by recycling the stuff of life into living organisms, they are also present at the beginning. And they don't grow old and die the way we do. A fungus can theoretically live forever as long as it has food to grow on, which is why one of the oldest and largest organisms on Earth is a fungus that lives on the top of a mountain in Oregon.[2] But keep in mind, a fungus can only be as big as its food source and as old as that source as well.

🍄 BIOLOGICAL CONNECTORS 🍄

Among the mushroom-producing species of fungi, there are three main lifestyles based on the way the fungi feed. They are kind of like culinary categories. Those lifestyle groups are the saprobic fungi (the decomposers), the mycorrhizal and endophytic fungi (the mutualists), and the parasites.

The saprobic fungi are the ones you see on stumps in the woods. They break down fallen trees, leaves, and so on and harvest their nutrients. If it weren't for saprobic fungi, we'd be living under miles of plant debris.

Of the mutualists, mycorrhizal fungi live on or in the roots of most plants where they exchange water and nutrients, like phosphorous, for sugar, which plants make through photosynthesis. It's a symbiotic relationship: Both plant and fungus cooperate to get the nutrients they need but can't make for themselves. Most plants have mycorrhizal partners, and some, like conifers, are utterly dependent on their root fungi for nonphotosynthetic nutrients. (If you squeeze the root tips of conifers, they will smell like mushrooms.)

This underground network of mycorrhizal fungi stabilizes soils as well. The fungus emits a sticky stuff called glomalin that glues soil particles together and waterproofs the particles so they don't blow apart when water passes by. Mycorrhizal fungi also function as a communication pathway that allows connected plants to share chemical information—for example, a chemical warning that aphids are on the move—with each other.

The other kind of mutualists are the endophytic fungi; that is, fungi that live between the cells of plants. Every plant that has been tested has endophytic fungi living between its cells.[3] So far, scientists have discovered that some endophytes provide defense services for plants by warding off predators and others provide stress tolerance, helping plants make it through tough conditions like abnormally high temperatures.

If a pony grazes on certain species of fescue grass, for example, it could get sick, but not because of the grass. They get sick because of the fungus that lives between the cells of the grass. The fungus is trying to ensure that the pony doesn't eat up its habitat. And when a plant is stressed, like during drought, it may produce particles that harm its own cells. Some endophytic fungi relieve that stress response by producing antioxidants that mitigate the damage to the plant's cells. This is an exciting area of research, because it turns out that endophytes that help one kind of plant deal with, say, high temperatures can be introduced into other kinds of plants so they can deal with high temperatures, too. And that's potentially prescriptive for a warming planet.

Much has been said and written about the role of mycorrhizal networks in forests, but these networks are just as important to farming. In the same way that fungal networks provide nutrients to trees, fungi in agricultural environments do the same for plants in a field or garden. But if you till a field, you tear up those fungal networks, undermining the plant's ability to get water and nutrients otherwise provided by the fungi. When you fertilize, the plant doesn't need to get necessary nutrients like phosphorus from the fungus anymore, so it stops paying out sugar and the fungus recedes. And fungicides spread on farmland work a lot like antibiotics work on us: There's collateral damage. You may be trying to kill a pathogenic fungi—because it's true that some fungi are

OPPOSITE *Mixed wild mushrooms.*

antagonists of plants—but you also may kill helpful fungi. So if there's a drought or some other environmental threat to the crop, the fungal network that's torn up by tilling or undermined by fungicide won't be there to back up the plant or stabilize the soil. That's one of the reasons why no-till agriculture—which replaces mechanical plowing with methods less disruptive to the soil—is recommended by the USDA.[4]

Finally, the third lifestyle group—the fungal parasites—are known to most farmers and gardeners because they prey on plants. The most destructive ones are capable of wiping out entire forests. The East Coast of the United States used to be home to huge forests of giant chestnut trees, the "redwoods of the East." They are all gone now, felled by a fungus.

🍄 A FOODIE PARADISE 🍄

We don't eat many parasitic fungi as food. The only one I can think of is *huitlacoche*, the tumors or galls on corn caused by a certain species of fungi. We mainly eat saprobic fungi. When you go to the grocery store and see a display of mushrooms, they are probably all saprobes—mushrooms that are the fruiting bodies of fungi that decompose for a living. That's because saprobic mushrooms are fairly easy to cultivate. You just need to give them their preferred food and appropriate atmospheric conditions. The most common cultivated mushroom is the *Agaricus bisporus*, which includes the white button mushroom, the cremini, and the portobello (they're all the same species). The cremini is just a brown version, and the portobello is a mature cremini. The flavor difference is negligible, though the portobello does have some extra taste because it has spores, and spores have flavor. Mushrooms on restaurant menus identified as "wild," such as shiitake and *Grifola frondosa*—also called maitake or hen-of-the-woods—are more likely to be cultivated than truly wild, though wild varieties exist. But if you see maitake on a menu in June in Los Angeles, you can be pretty sure it is cultivated—in the USA, wild maitake is a fall season, primarily Northeastern or mid-Atlantic mushroom.

The more expensive wild mushrooms are mutualists. Mushrooms like porcini and chanterelles are the fruiting bodies of mycorrhizal fungi that live in a symbiotic relationship with plants, mainly trees. They are difficult to cultivate because you'd have to plant an orchard and establish a soil microbiome to support the fungi. You'd have to reproduce the symbiosis among different living organisms. In the case of most mushrooms, that's just not financially feasible. As a result, mushrooms like chanterelles have to be found, hauled out of the woods, and then sold to a distributor before they finally end up in your risotto, so they are expensive.

The only mycorrhizal fungus we cultivate is the truffle. Because truffles cost so much, some folks are willing to take a chance and try to grow them. And to an extent, various types of truffles have been grown quite effectively. There are hundreds of truffle species, but only a few that we like to eat. Truffles are mushrooms that evolved to grow underground, but when they did that, they gave up their primary means of spore dispersal: wind. So truffles evolved to emit aromatic compounds when their spores are mature, which attract different animals depending on the species of truffle. Some species attract squirrels, but the truffles we prefer evolved to attract swine.

The famous white truffle of Italy has never been successfully cultivated. Maybe it's because another symbiont is necessary to make the fungus fruit. Certainly, there are other symbionts at play underground. Indeed, bacteria living on the truffle may be responsible for the fabulous aromatics we are so in love with in the first place. Anyway, we tend to think of symbiosis as having two players, but in nature, everything is symbiotic to some degree.

Humans can detect five main types of flavor: sweet, salty, sour, bitter, and umami, a meaty savoriness. Mushrooms are umami-flavored, because they contain all nine amino acids needed to make a complete protein—not as many as steak but still high quality.[5] That protein is the reason why, if you substitute mushrooms for meat in your recipes, you will take in fewer calories but still feel full. Mushrooms are also a source of vitamin D_2. When mushrooms are irradiated—something like putting them on a tanning bed—the ultraviolet light converts a compound in them into vitamin D_2, not unlike the way sunlight turns a compound in our skin into vitamin D_3.[6] High-grade protein and the ability to make vitamin D are just two of a number of similarities humans share with the fungi kingdom.

🍄 MEDICINE FROM THE KINGDOM 🍄

There are medicinal mushrooms and medical mushrooms. Medical mushrooms (or, more accurately, medical fungi) affect our bodies in some negative way, like toenail fungus. If you've had one, you know how difficult it is to get rid of a fungal infection. That's because killing fungi that live on us is not so easy. On the tree of life, fungi are closer to animals than to plants, so the medicine that kills a fungus can hurt us as well. Doctors learn about medical mushrooms in school, because we all commonly deal with fungal infections.

Medicinal mushrooms, on the other hand, are used as curatives. For example, some fungi produce compounds that act like chemical warfare agents: They fight off competitors or intruders, such as bacteria, viruses, and other fungi. The penicillium mold, which is the source of penicillin, is one such fungus commonly used in Western medicine. Traditional Chinese medicine and other naturopathic practices, however, prescribe a variety of mushrooms and fungi in the form of extracts, teas, and foods that are less about attacking disease (like penicillin does) and more about supporting health. For example, *Ophiocordycep sinensis*—also known as the caterpillar mushroom—is a highly prized medicinal mushroom that is a parasite of Tibetan ghost moth larvae. The mushroom is used for lots of ailments but is considered particularly helpful if you are trying to keep secondary infections at bay, so if you are recovering from a catastrophic illness and your immune system is compromised, the mushroom may help you stay well while you heal.

ABOVE *Cooking with mushrooms.*

The ghost moth larva is a small orange worm that becomes infected with an *O. sinensis* spore. The spore germinates and grows, eating the worm from the inside and ultimately killing it, all the while producing chemicals to ward off competitors for the larva carcass, like other fungi and bacteria. Eventually, the fungus produces a fruit body out of the head of the larva. It's not really a mushroom, but it plays the same role in the reproductive cycle of the fungus. Anyway, the medicine is not in the mushroom; it's believed to be in the worm, which the fungus has riddled with defensive antibiotics. Again, because we are close to fungi on the tree of life, we may benefit from those antibiotics as well.

🍄 MICROBIAL WISDOM 🍄

The culture of mycophilia—the love of mushrooms—seems to be growing. It's not just about eating mushrooms or even hunting them. More and more people seem to be interested in the biology of these organisms, their ability to support plants both nutritionally and defensively, and their role in human health. It's a revelation to realize that charismatic life-forms like trees have microscopic partners that play key roles in their well-being. It's like the news we keep hearing about the bacterial symbionts in our guts. The unseen world supports the seen world, and yet, it was only recently that we discovered they were there.

For me, mycology was a gateway science to the incredibly complex world of microbiology. Fungi are microscopic and mushrooms are macroscopic, so the study of mycology bridges the unseen and seen worlds. It's hard to comprehend microscopic life, but we're beginning to recognize that every organism you can see is an environment for an ecology you can't see. Organisms like bacteria and

fungi live in complex microscopic communities that actually function according to the same rules of ecology as the animals in Yellowstone Park. So in a sense, if you understand a forest, you can understand a teaspoon of soil.

Ecology—who eats who and who helps who eat who—is what supports the insights of mycologist Paul Stamets and others, allowing them to understand how, for example, fungi pesticides might work or how fungi might remediate soil. But if you aren't looking at microscopic life from an ecological point of view, it's pretty hard to see anything. And it's important that we do see, because as the great microbiologist Moselio Schaechter once told me, "Eugenia, half the world is microbial. Without a little microbiology, you can't understand half of yourself."

NOTES

1. David L. Hawksworth and Robert Lücking, "Fungal Diversity Revisited: 2.2 to 3.8 Million Species," *Microbiology Spectrum* 5, no. 4 (2017), http://www.asmscience.org/content/journal/microbiolspec/10.1128/microbiolspec.FUNK-0052-2016.
2. Nic Fleming, "The Largest Living Thing on Earth Is a Humongous Fungus," BBC.com, November 14, 2014, http://www.bbc.com/earthstory/20141114-the-biggest-organism-in-the-world.
3. M. Jia, L. Chen, H-L. Xin, et al., "A Friendly Relationship between Endophytic Fungi and Medicinal Plants: A Systematic Review," *Frontiers in Microbiology* 7 (2016): 906, https://www.ncbi.nlm.nih.gov/pmc/articles/PMC4899461/.
4. Christy Morgan, "No-Till Leads to Healthy Soil and Healthy Soil Leads to a Better Growing Season," USDA Natural Resources Conservation Service, www.nrcs.usda.gov, accessed July 2018.
5. Z. Bano, K. S. Srinivasan, and H. C. Srivastava, "Amino Acid Composition of the Protein from a Mushroom (*Pleurotus* sp.)," *Applied Microbiology* 11, no. 3 (1963): 184-7.
6. "Mushrooms and Vitamin D," BerkeleyWellness.com, December 5, 2016, accessed July 2018.

ABOVE *Oyster mushrooms* (Pleurotus ostreatus). OPPOSITE *Mixed wild mushrooms.*

🍄 DELICIOUS MUSHROOM RECIPES 🍄

A few words about buying, storing, and cooking mushrooms. Mushrooms are the fruiting bodies of fungi, and as such, they should be purchased and stored much the same way as you would flowers or fruit. When buying mushrooms, look for the elasticity of youth in the texture and a lovely earthy smell. Like berries, do not wash mushrooms until you are ready to prepare them. Keep them unwashed in a paper bag in the fridge. Mushrooms can be sauteed, roasted, broiled, grilled, or boiled. Some species can be eaten raw, but not all—morels, for example, will make you sick if you eat them raw. When in doubt, cook mushrooms. Here are a couple of favorite recipes:

PENNE AI FUNGHI WITH ARUGULA
Serves 4

INGREDIENTS

2 cups mixed organic mushrooms—for example, oyster mushrooms, maitake, or shitake—coarsely chopped with tough stems removed
5 tablespoons olive oil, divided
¼ cup warm chicken (or mushroom or vegetable) stock
1 tablespoon garlic, minced
1 tablespoon plus 1 teaspoon lemon juice, divided
1 teaspoon dried marjoram
Salt and freshly ground black pepper to taste
¾ pound penne pasta
2 cups fresh arugula, washed and torn into bite-size pieces
Grated Parmesan cheese (optional)
Flat-leafed parsley, chopped (optional)

This recipe is based on the idea of pasta fagioli. It works surprisingly well. You can substitute the arugula with watercress.

Preheat your oven to 450°F. Place the mushrooms on a cookie tray and toss with 2 tablespoons of the olive oil. Roast for about 20 minutes, shaking the pan once or twice to turn the mushrooms, until they are fork tender. Place the mushrooms and the stock in a food processor and pulse to a grainy puree.

In a large pan, heat 2 tablespoons of the olive oil over medium heat. Add the garlic, 1 tablespoon lemon juice, and marjoram and cook for a few minutes until the garlic becomes aromatic but not browned. Add the mushroom puree. Cook long enough to heat through.

Add salt and pepper to taste.

In the meantime, bring a large pot of salted water to a boil and add the penne. Cook until al dente, about 12 minutes. Drain and add the pasta to the mushroom sauce. Stir to combine mushroom sauce and penne and heat over medium-low heat.

Toss the arugula with the remaining tablespoon of olive oil and 1 teaspoon of lemon juice. You can adjust the oil or lemon juice to your taste. (You don't have to dress the salad at all—it's very good plain as well.)

Serve the penne with 1/4-cup garnish of arugula on each plate.

As an alternative, dress the pasta with grated Parmesan cheese and chopped flat-leafed parsley, to taste.

WILD MUSHROOM SOUP

Serves 4

INGREDIENTS

2 tablespoons unsalted butter

1 cup onion, minced (about ½ large onion)

1 pound wild mushrooms—for example, porcini, chanterelles, hedgehogs, or maitake—sliced

⅓ cup sweet Marsala or Madeira

1 tablespoon flour

4 cups chicken (or mushroom or vegetable) stock

2 sprigs fresh thyme

4 tablespoons mascarpone cheese (or heavy cream)

Chopped fresh thyme for garnish

Salt and freshly ground black pepper

To get that wild taste without purchasing costly fresh mushrooms, replace the wild mushrooms with 5 ounces of dried porcini mushrooms rehydrated in 5 cups of warm water for about 20 minutes until soft and 1 pound of sliced white mushrooms. The dried porcinis add the flavor, and the buttons add the texture.

Melt the butter in a heavy soup pot over medium heat. Add the onions and cook them until they are soft, about 3 minutes. Add the mushrooms and sauté them until they give up their liquid, about 15 minutes. Add the Marsala wine, cover, and bring to a boil. Remove the cover and allow the wine to cook out, about 3 to 5 minutes. Stir the flour into the mushroom mixture. Add the stock and thyme sprigs. Bring the soup to a boil, then turn down the heat and simmer for about 20 minutes, stirring occasionally, until the flavors meld.

Remove the thyme sprigs. Remove about half the mushrooms and grind them in a food processor. Return the ground mushrooms to the soup and combine. The soup should be about the consistency of corn chowder. If it seems too thick, add some more stock or warm water.

To serve, swirl a tablespoon of mascarpone into each bowl of soup and garnish with chopped thyme.

🍄 FUNGI FACTS:

THE FIRST CULTIVATED MUSHROOM

ABOVE *White button mushrooms* (Agaricus bisporus). **OPPOSITE** *Unidentified mushroom species.*

The first mushroom to be cultivated in the United States was the *Agaricus bisporus* (commonly known as the white button mushroom), but we have France to thank for that. During the reign of Louis XIV, French cultivators noticed the mushroom growing on horse manure. They transposed the mycelium underneath it to compost beds, where the mushrooms would grow in controlled abundance. They eventually realized they could dry the mycelium and ship it. That's one of the really interesting things about mycelium: It can dry out and lay dormant for a long time, but as soon as you add water, it wakes up.

There was a problem, though. This dried mycelium was basically a wild product, and it contained all kinds of microbial hitchhikers. When American buyers purchased and cultivated it, other fungi and bacteria would sometimes kill it. This motivated the USDA to figure out how to grow *Agaricus bisporus* from spores, thereby ensuring both the species and the purity of the culture. Virtually all white button mushrooms today originate from the same spore called the U1 hybrid, which was discovered by a Dutch scientist named Gerda Fritsche in 1980. This "superspore" was selected and continues to be preferred because it produces a high yield of disease-resistant mushrooms with good flavor. Indeed, one piece of *A. biosporus* mycelium, the size of a watch battery and the weight of a housefly, can produce 100,000 pounds of white button mushrooms.

CHAPTER 12

A FORAGER'S HOMAGE

༄ ༄ ༄

GARY LINCOFF

Though some refer to mushrooms and fungi as the third kingdom, to me they are the magic kingdom. The learning is endless, and no matter where I am, they find me.

Mushrooms energize me. I take them like some people take vitamins. I only need to think about them to start feeling the excitement.

Among all biologists, I adored the late Lynn Margulis, and I read everything she ever wrote. She once said that we couldn't understand mycorrhizae, we couldn't understand symbiosis, and we couldn't understand mycelial connectedness, because our society is trapped in a capitalist, individualistic mindset. Anything that smacked of communalism was communism. Even if you were a scientist, if you were sympathetic to a bigger, more inclusive perspective of what makes the world tick, you couldn't be trusted. For many generations, we've suffered from the myth that individuals work better alone. But when it comes to mycelium, there is no such thing as alone.

🍄 ENDLESS LEARNING 🍄

I live in New York City, and when I first started learning about mushrooms in the 1970s, I discovered that no one in this country knew a thing about them. Even at the New York Botanical Garden they didn't know what mushrooms were—and still don't. It's not a mainstream thing.

One day years ago, I went for a walk with a group from one of those nerdy botanical societies where they couldn't care less about the nutritional or medicinal benefits of a plant; they just wanted to name it. I happened to be carrying a plastic bag of inky cap mushrooms. I can't remember why. An hour into the walk, the bottom of the bag was filled with black muck. Three of the people on the walk looked at the bag and said, "Oh, you've got inky caps." And I said, "Whoa, how do you know that?" And they said, "Well, we're members of the New York Mycological Society." I was stunned. I'd been looking for them for some time. They weren't in the phone book. No one knew who they were. "Oh! I've been looking for you!" It was like finding a rare mushroom.

The more time we spent together, the more we realized how little we knew about mushrooms—so we decided to take a trip. We first went to China and studied mushrooms there for three weeks, spending a lot of time with the locals. We went to Japan and did the same thing. We ended up traveling to thirty different countries across six continents meeting people and studying mushrooms. And we kept asking them questions like, "What do you have? Where do they grow? How many are like ours? How are they different? What are your rituals?" We found entire groups of people who had ideas totally different from ours, and even from each others'. Russians, for example, love mushrooms.

OPPOSITE Lycoperdon perlatum, *popularly known as the common puffball.*

Everyone knows that. And so do the Koryaks, an ethnic minority that lives in Far Eastern Russia. But the Koryaks won't eat Russian mushrooms, and the Russians don't trust the Koryaks. Case in point: The Koryaks have a strong relationship with *Amanita muscaria*, known for its strong hallucinogenic qualities. The Russians believe it will kill you.

That's one of the things I love about mushrooms. The learning is endless, but you are always making progress. You know more than you knew last year, and you know much more than you did five years ago. You correct your mistakes as you learn along the way, because we're all making mistakes. We're all misidentifying things. So what? It's all about the exploration.

🍄 URBAN FORAGING 🍄

"We've been systematically visiting city parks every week since 2011, and we've found almost a thousand different mushrooms. A thousand. In New York City!"

I live within walking distance of Times Square, and just another five minutes from Central Park, which is eight hundred acres. Nearly every weekend—winter included—I lead a group of my colleagues from the New York Mycological Society, along with a few others, all of us amateurs, to one of the city parks to collect and survey the mushrooms we find and try to discern patterns to what we're seeing. I've got about fifteen confirmed addicts who, no matter the weather, will come out and hunt mushrooms, even on football Sunday. That really shows passion. And every single time we go out, we can guarantee finding fifty different mushrooms. Brooklyn's Prospect Park, at five hundred acres, is fabulous for mushrooms. Queens is full of parks and also really good for mushrooms. We've been systematically visiting city parks every week since 2011, and we've found almost a thousand different mushrooms. A thousand. In New York City!

I think we've transformed the notion of urbanity and urban living into a larger, natural matrix that says we coexist with nature no matter how much we believe we don't. The buildings say we don't and the shops say we don't, but we do.

I wouldn't choose to, but even if I lived in a desert I'd still be a mycologist, and my first mission would be to find desert truffles. It's hard to explain, but mushrooms are in some ways more real to me than where I live.

SPOTLIGHT ON:
GARY LINCOFF: MYCO-VISIONARY (1942–2018)

Gary Lincoff was a true original: a passionate, enthusiastic advocate for the interconnectedness of nature and a beloved figure in the field of mycology. He wrote a number of mushroom identification guidebooks including the seminal *National Audubon Society Field Guide to North American Mushrooms* (1981) and, most recently, a revised edition of *The Complete Mushroom Hunter: An Illustrated Guide to Foraging, Harvesting, and Enjoying Wild Mushrooms* (2010), both considered essential reading for the amateur and professional alike. He was an early member of the New York Mycological Society, he had served as president of the North American Mycological Association (NAMA), and he was active with the Connecticut-Westchester Mycological Association (COMA), which described him as "the Socrates of mycology" because of his deep and thoughtful approach to educating the public. He also taught classes at the New York Botanical Garden for forty-two years and educated over four thousand students in over three hundred and fifty courses. His heart and his influence will long be remembered.

ABOVE *Gary Lincoff.*
OPPOSITE *Mixed wild mushrooms.*

CHAPTER 13
A MUSHROOMING OF INTEREST

BRITT BUNYARD

BRITT BUNYARD is publisher and editor in chief of *Fungi Magazine*, a college professor with a PhD in plant pathology from Pennsylvania State, and a practicing mycologist. He was also editor in chief of *McIlvainea: Journal of American Amateur Mycology* and subject editor for the *Annals of the Entomological Society of America*.

A growing and diverse community of researchers, foragers, and citizen scientists has transformed the world of mycology. It's the "new thing," but like ecotourism, we need to remain good stewards.

We started *Fungi Magazine* twelve years ago to fill a gap between academic journals, which tend to be dry and technical, and niche newsletters with small readerships and limited coverage. There are a lot of enthusiastic mycophiles out there who are foraging or into mushroom photography or have some other interest in the various applications of fungi and mushrooms but weren't getting the information they needed from any other source, so the magazine has really taken off. The Telluride Mushroom Festival, which started back in 1981 as Wild Mushroom Telluride, now mirrors the diversity of the magazine. There are a lot of different presenters and topics, and while they may not be academic or professional mycologists, they are the real experts.

Normally we sell about one hundred full passes and then hundreds of various single-day passes, but in 2018 we sold out almost immediately—all the full passes and all the open-to-the-public events. Why? Well, interest has been steadily increasing, but a few things were different this year. One was that we featured Paul Stamets, who always draws a crowd and who is the longtime heart and soul of this event. Another was the memorial we put together for Gary Lincoff, who was beloved by many. And I think that Michael Pollan's new book *How to Change Your Mind* had an impact because of its focus on the therapeutic benefits of psilocybin, which is produced by more than two hundred species of mushrooms. So it was a perfect storm of all that.

ABOVE *The panther cap* (Amanita pantherina *group*). OPPOSITE *Little brown mushrooms, species unknown.*

ELINOAR SHAVIT is an ethnomycologist. She is a past-president of the New York Mycological Society and chair of the Medicinal Mushrooms Committee at the North American Mycological Association (NAMA). She is an author and frequent speaker in the United States and abroad on the topics of medicinal mushrooms, desert truffles, and ethnomycology.

ABOVE *White truffle.*
OPPOSITE *Black truffle.*

SPOTLIGHT ON: DESERT TRUFFLES

In the vast deserts of the Middle East and southern Mediterranean grows a family of truffles collectively referred to as desert truffles. These desert truffles are an important foodstuff for indigenous dwellers in these deserts, as they produce highly nutritious fruiting bodies after it rains, usually at a time when most stored provisions have been used but fresh ones are not yet available. Travelers to Arabia reported that some Bedouin tribes would eat nothing but desert truffles for the entire desert truffle season.

Desert truffles are cherished wherever they grow and have the longest recorded history of continuous use by humans of all the mushrooms. Their history practically follows the history of human civilization. They are discussed in the ancient records of the Sumerians, Egyptians, Greeks, and Romans and also are mentioned in the Jewish Talmud and Islam's religious writings. There is even merit to the claim that desert truffles are the biblical manna. The original Hebrew texts in the Bible regarding manna tell the reader what the white *Tirmania nivea* desert truffle looks like, where and when to find it, how to gather it, and even how to cook and preserve it. These biblical passages read like the *Tirmania nivea* page in a modern field guide to desert truffles, and those instructions—for the most part—are still followed today by Bedouins and other desert truffle gatherers.

Regrettably, climate change and the encroachment of human civilization on the traditional growing areas of desert truffles threaten the survival of the truffles' habitat as well as the culture and way of life of the indigenous people who celebrate them.

—ELINOAR SHAVIT

✦ A DIVERSE COMMUNITY ✦

The mushroom community is made up of different threads. There are those who relate to mushrooms as mindblowers, those who love to forage, and those who love to cook with them. Then you've got researchers at places like UCLA and NYU who are using psilocybin as therapeutic aids. It's quite a diverse culture.

"People like you and me are just doing it themselves, learning how to forage and research and experiment on their own."

Historically, interest in mushrooms was mostly culinary, along with some scientific interest and artists drawn to their startling imagery, but that was about it. The counterculture brought the psychedelic component to the surface, and now each of those groups has been growing and morphing. Within culinary circles, for example, certain mushrooms, such as shiitake, oyster, chanterelles, and morels, are suddenly cool and hip and everyone wants in on them.

Foraging in general, for plants, herbs, and other wild things, is on the upswing. But the idea of growing and harvesting mushrooms for their medicinal properties, converting them into teas and tinctures and the like—that wasn't happening ten or twenty years ago. That particular area has exploded. And now—again, thanks to Paul Stamets—people are starting to use fungi and mushrooms to clean the planet, applying them to oil spills and toxic waste sites. This mycoremediation has huge potential.

All of this has been driving interest in mycology, and it's not the academics. People like you and me are just doing it themselves, learning how to forage and research and experiment on their own. The field has really benefitted from these citizen scientists.

So, there's a lot of momentum. Mycelium and mushrooms and fungi are becoming the new "new thing." In 2018, there were attempts to get a measure to legalize psychedelic mushrooms on the Colorado ballot. I thought, "You've got to be crazy. That's never going to happen." And then I remembered, "Oh, wait a minute. I vividly recall telling people ten years ago that cannabis would never be legal."

ABOVE *Lentinus strigosus.* OPPOSITE *Shelf fungi growing on a hardwood tree in Necedah National Wildlife Refuge in Wisconsin.*

🍄 WHAT'S NEXT? 🍄

There are definitely some growing pains. Wild mushrooming is becoming a little too popular in some areas of the country. In a lot of the Western states, for example, especially in the national forests, you need a permit to go mushrooming. In Washington, collection regulations can be as specific as which fire or ranger district you're in, and there are limits on how many of what kind of mushrooms you can take. It depends on the mushrooms; if the variety is commercially important, you'll need to have a permit and pay a collecting fee. So everyone is out beating the woods, looking for mushrooms. It's a bit like ecotourism. The danger in a few places might be in loving it to death and not being good stewards of the land. And, of course, knowing which ones are edible and which ones aren't.

But overall, I believe these developments are very positive, especially regarding the potential of these agents to clean up our planet. It's already happening. You hear a lot of doom and gloom stories about lost habitats, water pollution, and so on. But there are also good news stories out there about communities that have realized they can make a difference.

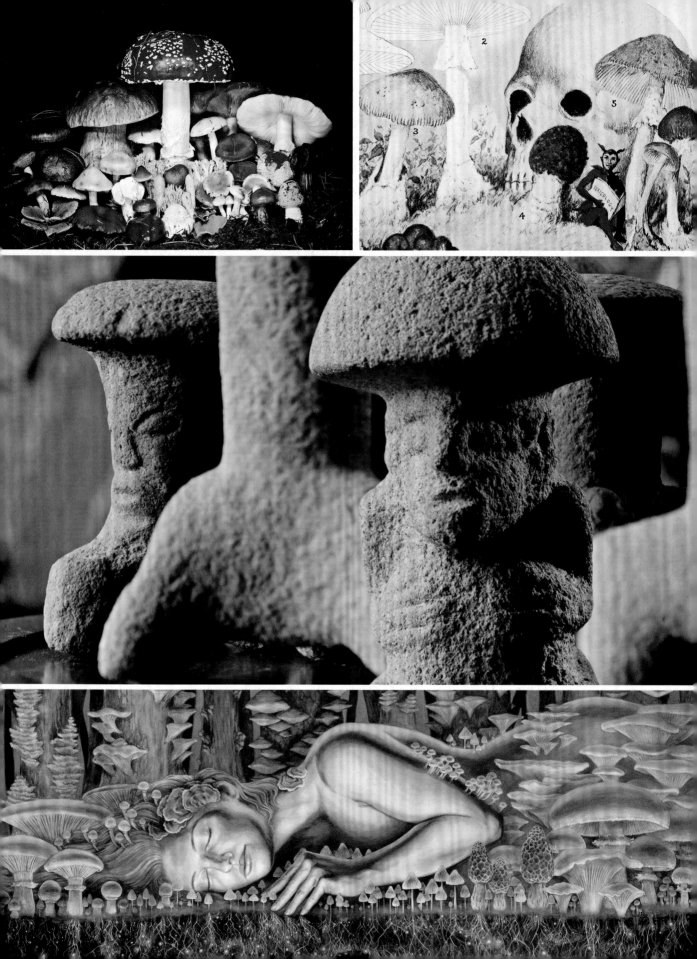

FUNGI FACTS:
MUSHROOMS ACROSS TIME: AN ANCIENT HISTORY

Mushrooms, whether for nutrition, medical benefit, or ecstatic experience, have been used by different cultures throughout the world for thousands of years.

- Mushroom spores were found in the remains of a woman who died nearly twenty thousand years ago in what is now Spain. Researchers speculate that they could have been nutritional, medicinal, or even ceremonial.
- Mushrooms were referred to as "sons of the gods" in Egyptian times; hieroglyphics show Egyptian nobles dancing around enlarged images of what appear to be mushrooms. Other Egyptian texts highlighted recipes with mushrooms as a main course for its nutritional value.
- The mummified body of the iceman Ötzi, so named because it was found in the Ötztal Alps along the border separating Austria and Italy, dates to approximately 3300 BCE. He was carrying two types of polypores: the birch (*Piptoporus betulinus*), which grows almost exclusively on birch trees and is still used to fight infections and as a digestive aid, and the tinder fungus (*Fomes fomentarius*), well known as a fire starter and for its ability to carry embers long distances.
- They have a long history in Asia: References to reishi (lingzhi or *Ganoderma lucidum*) have been found in two-thousand-year-old writings. Buddhist monks traveling from monastery to monastery were said to spread information about the curative effects of fungi, which they and Taoist priests used in rituals.
- The ancient Greeks and Romans considered mushrooms a mystery, as the organisms were neither plant nor animal, and they had no discernible source. These ancient peoples treated mushrooms as delicacies for the privileged and later recognized mushrooms for their potential to increase health and stamina.
- Varieties of the chaga fungus (*Inonotus obliquus*) were used across Siberia, Scandinavia, and other northern-latitude regions for a range of ailments having to do with infection, inflammation, stamina, pain relief, and what we would now describe as cancer.
- Archaeologists discovered statues that depicted what they called "mushroom shamans" in Mexico and Guatemala dating back thousands of years. The Mayan, Toltec, and Aztec cultures of Mexico ritualized the consumption of a broad range of entheogens, including psilocybin (the "flesh of the gods").

—ELINOAR SHAVIT

OPPOSITE TOP LEFT *An assortment of mushrooms (© Taylor Lockwood).*

OPPOSITE TOP RIGHT *Artwork depicting poisonous mushrooms.*

OPPOSITE MIDDLE *A Mayan mushroom stone that is over two thousand years old.*

OPPOSITE BOTTOM *Mycelium Dreaming © Autumn Skye Morrison.*

CHAPTER 14

GROWING YOUR OWN

֍ ֍ ֍

KRIS HOLSTROM

KRIS HOLSTROM is the owner of Tomten Farm, a high-altitude, off-grid, permaculture-based research, demonstration, and education farm near Telluride, Colorado. She consults on permaculture designs, edible and medicinal landscaping, and school and community gardens through the nonprofit SWIRL, the Southwest Institute for Resilience.

One of the great things about mushrooms is that they're easy to grow and we can learn from our mistakes. Decide which ones you want, do a bit of research, and you'll be ready to go!

I was introduced to the possibilities of growing mushrooms at the Telluride Mushroom Festival by Andrew Weil and Paul Stamets many years ago, and I continue to be inspired by the new voices of Tradd Cotter, Daniel Reyes, Leif Olson, and many others. It doesn't take much for the mushroom bug to bite and take hold if you are at all concerned about the environment, your health, or the future of our planet.

Thorough and fantastic books are available for diving into mushroom cultivation, but the best advice I've heard is: Don't be intimidated or delay your foray into mushroom growing because it seems too complex. Just get growing!

Consider:

1. What kind of mushrooms do you like? We're fortunate to have several great suppliers of mushroom spawn in the United States.

2. What kind of resources for a substrate—the material that the mycelium and mushrooms use to grow on—do you have readily available? Wood chips, straw, cardboard, hardwood logs?

Being primary decomposers, many of the most delectable and easy-to-grow mushrooms prefer hardwoods like oak or alder. Oyster mushrooms seem to be the omnivores of the cultivation world and can grow on everything from cardboard to coffee grounds to straw.

Our journey at our permaculture farm, Tomten Farm, started with those two questions. We have lots of Gambel oak, and we enjoy shiitake mushrooms. A little research and some tips from those with experience led me to do my first oak log inoculation using a "plug spawn"—wooden dowels that come already myceliated and ready to go. We cut three-to-four-inch-diameter oaks into convenient lengths. After about two weeks (the pieces need some time to mellow so the sap settles), we got busy. A drill with a 5/16 bit, a rubber mallet, and some beeswax are the basic necessities.

We first drilled a herringbone pattern of holes covering the entire log. Special drill bits are available that stop at the right depth, or you can mark the bit and watch for the depth you want as you

OPPOSITE Armillaria *sp. growing on a mossy stump.*

drill. You want to avoid air pockets at the bottom of the hole. The next step was pounding the dowels into the holes with a mallet. The final step was to cover the holes with beeswax to seal in moisture and seal out competing spores.

We stored our logs behind a shed so they would be covered with snow all winter. The next spring, we moved them to a shady area where we could tend them, watering on occasion in our very dry climate. It took a year for our first shiitakes to emerge. The second summer we had significant fruiting after soaking the logs in a barrel of water for a few days. Depending on your conditions (we had exceptional drought and had to compensate), you could see fruiting within six to twelve months.

I also have friends in the tree business who are always looking for places to deposit their wood-chip waste. I can get mixed conifer and hardwood chips anytime, and if I ask real nice, they'll bring up a load of pure hardwood. This is usually free, but I'll always give them a few dollars for their time and gas.

For this particular wood-chip substrate, we made a simple outdoor bed using untreated scrap lumber. The chips were spread in the bed with each layer getting water and a generous sprinkling of stropharia (the garden giant) sawdust spawn from Mushroom Mountain. We covered the bed with some garden row cover early on, as we were experiencing a crazy hot and dry June and wanted to protect the spawn from the heat. By the end of summer—ours are short—the bed was fully myceliated with just the top inch dry. We started to get fruiting in the early fall, then covered the bed with insulating straw and will see what happens next year.

We also built a mushroom room in our greenhouse to experiment with turning one of my least favorite plants—thistle—into substrate. And it worked! My friend pulled a truckload of thistle from his property, and with his financial help and the help of a farm intern, we experimented. First, we dried and shredded the thistle stalks. We tried four combinations of thistle and straw, pasteurizing (which kills most of the competing spores and microorganisms) different batches using heat, chemical, and anaerobic means to get the substrate ready for inoculation (the process of mixing substrate with mushroom spawn). We then packed clear plastic bags with the substrate and oyster spawn. Over the next few weeks and months we measured production and were delighted to discover that the 100 percent thistle substrate had the highest production. These kinds of surprises are a big part of what makes mushroom cultivation so enjoyable.

If even these simple methods seem overwhelming, consider starting out by purchasing a mushroom kit for your kitchen counter. These are great gifts (perfect for students) that don't require a lot of personal involvement or depend on the use of local waste products.

Playing around is a great way to grow and learn to love mushrooms. They are amazing in their variety and their ability, as we are beginning to discover, to decompose our "waste" products. Think of the amounts of cardboard, untreated wood products, and other organic materials that are landfilled. Much of that could be used for growing mushrooms. Waste becomes resource. As you gain more experience, you may want to try out more sophisticated techniques using flow hoods and controlled environments for production. It's easy and oh so satisfying to convert what would be wasted into a mycelial process and then delight in the accomplishment of growing your own mushroom feast.

FUNGI FACTS: EATING WILD

KATRINA BLAIR is a wild-plant advocate and author. Her books include *Local Wild Life: Turtle Lake Refuge Recipes for Living Deep* and *The Wild Wisdom of Weeds: 13 Essential Plants for Human Survival.*

ABOVE *Mixed wild mushrooms.*

When we eat directly from the land we live on, those plants and mushrooms resonate with an intelligence that is brilliant. This interaction reminds us that we are also brilliant. Because we are what we eat, we become connected deeply to the earth around us. When we put local wild food into our bodies, our intuition becomes heightened and we are reminded to be good stewards of the planet. The roots of dandelions, for example, go deep down into the soil to pull up minerals and store them in their leaves, which then mineralize us when we eat them or drink their tea. Mushrooms are about breaking things down and recycling them. When we eat them, they help break down the things inside us we don't need. They make our entire body more efficient. They help in our constant renewal, resiliency, and ability to thrive.

—KATRINA BLAIR

CHAPTER 15
MUSHROOMS TO THE PEOPLE

WILLIAM PADILLA-BROWN

WILLIAM PADILLA-BROWN is a social entrepreneur, citizen scientist, mycologist, amateur phycologist, urban shaman, poet, father, and founder of MycoSymbiotics LLC, a mycological research and mushroom-production business based in New Cumberland, Pennsylvania.

It's never too late to follow your dreams. My own journey toward becoming a mycologist and community organizer had its ups and downs, but I wouldn't trade it for the world.

I dropped out of high school when I was sixteen because it was interfering with my education. I realized at a young age that a lot of people weren't learning what they really needed in order to live a successful life, so I started to seek an alternative lifestyle. Mushrooms and permaculture design science became the key to unlocking that potential for me.

I started by doing a lot of internet research and attending workshops and ultimately developing a nontraditional, independent education system for myself. I met a lot of gardeners and permaculture scientists in my area, and as I got more involved with them, I realized that most knew little about mushrooms, and yet every single one had mushrooms in their garden. So I researched some more and discovered that no one in a hundred-mile radius of where I lived had information on mushrooms or was teaching anybody about them.

Now I'm a mycologist, a mushroom farmer, a certified permaculture designer, and an educator. I focus a lot of my time on understanding patterns and then designing techniques to bring that knowledge into urban settings. I focus on teaching people in inner cities how they can use this information and these sciences to alleviate economic stress and build a better life for themselves. I go to farmers markets and work with children to increase their ecological literacy. I'm basically exposing these people to new realities, showing them that there are other ways to live that won't get them arrested. I see a lot of people who don't know where their food comes from, who don't know how they're connected to their community, who don't know how they're connected to the environment. In the inner city, most folks haven't had the opportunity to walk out in nature, and those that have often haven't had the education to understand what it is they're experiencing.

I also see a lot of food deserts in urbanized areas where there aren't a lot of grocery stores, and when there are stores, you see mostly packaged and processed foods—certainly not a lot of fresh food. Local folks have no relationship with the food they eat. So if I can show them real food, especially food that's grown right there in their community, they start to get it, they start to understand the holistic world they live in. Seeing healthy, nutritious food right in front of them—especially mushrooms, because they are so nutrient-dense—and knowing where that food comes from, that is the way to open people up to new possibilities that will have a very real and positive impact on their lives.

OPPOSITE FROM TOP LEFT TO BOTTOM RIGHT *Oyster mushrooms* (Pleurotus); *Cultivated* Cordyceps militaris; *Cultivated brown Agaricus bisporus*; *Oyster mushrooms* (Pleurotus ostreatus *group) fruiting from straw.*

🍄 BREAKING NEW GROUND 🍄

I implement a lot of my work through MycoSymbiotics, a mycological research and production business I started in 2015. We don't have a website per se, but we do have a comprehensive blog, podcasts, and a robust social media presence. Our home base in Lemoyne, Pennsylvania, just outside of Harrisburg, has a laboratory and a mushroom farm that we utilize for educational purposes. People can tour our facilities and get hands-on with the science there, but I also take everything on the road. A lot of people can't afford to get out of the city on a regular basis, so I teach at schools and various events to bring this information to places that otherwise would never get it. I'll also go to urban farms and inner-city permaculture farms like Baltimore's Charm City, where I teach or plant mushrooms and learn what others are doing.

"Mushroom 'farms' can then be built in barns and sheds and basements, all places that are easily accessible to most people. For a little bit of money you can get a very big yield."

I first started growing mushrooms in a closet, then expanded into my home, and became very passionate about cultivating gourmet and medicinal mushrooms in small spaces. And so my approach to mushroom farming is different from conventional methods that use multimillion-dollar laboratories and farming operations. I focus on developing low-tech techniques that anyone can apply. I know of other mycologists who are experimenting with similar techniques, and we share our knowledge and experience. Bringing this work and these mushrooms to more people reflects the spirit of what we're all doing. These new techniques utilize agricultural wastes and urban wastes like coffee grounds and cardboard to create low-cost environments for cultivating mushrooms. Mushroom "farms" can then be built in barns and sheds and basements, all places that are easily accessible to most people. For a little bit of money you can get a very big yield.

I also have a strong interest in mycomedicinal mushrooms. Both of my grandparents on my mother's side passed away from cancer; they were taking all sorts of pills and living very uncomfortably by the end of their lives. If only they'd known about some of these immunomodulators, like the reishi mushroom. I'm especially interested in the cordyceps fungi, specifically *Cordyceps militaris*, which is another powerful immune system booster.[1] People from all over the world send me specimens. I'm looking to create resilient commercial strains for an emerging microculture of cordyceps farmers. This sort of farming is very new to the United States. People have been farming cordyceps mycelium for medicinal products for some time, but the farming of the actual cordyceps fruit body was initiated by myself over the past two years. My goal is to increase its production and get it out to the masses.

In addition to that, I'm always seeking gourmet and medicinal cultures everywhere I go. If I find a fungal specimen that's doing something interesting, maybe growing in an odd location or playing some kind of mycoremediation role, breaking down something funky, I bring it back to the lab and study it. I share the information with my friends in this wonderful mycelial network that I've become a part of to see what they can figure out. Our objective is to identify different genetic barcodes and see if maybe we've found a new species or something to add to the developing phylogenic tree. I believe that DNA analysis will just keep getting bigger as an analytical tool.

OPPOSITE *William Padilla-Brown working with his community to cultivate an urban mushroom garden.*

🍄 FOLLOW YOUR BLISS 🍄

Many people around me thought mushrooms were weird, but that's where my passion went, and I followed it. The journey has been incredible and fulfilling. I feel like I'm really helping people in my community, feeding them and providing medicines that might heal them. My experience has taught me how important it is for individuals to follow the things they're interested in, regardless of what others say, especially when it's something that is ethical and ecological and that can help us all live more holistically on the planet. Do what you love, give it your best, and share it. The people who will support you may not be in your immediate community, but they're out there.

NOTE

1. Yong Sun et al., "Regulation of Human Cytokines by *Cordyceps militaris*," *Journal of Food and Drug Analysis* 22, no. 4: 463-67, https://www.sciencedirect.com/science/article/pii/S1021949814000301; M. Jeong et al., "Cordycepin-Enriched *Cordyceps militaris* Induces Immunomodulation and Tumor Growth Delay in Mouse-Derived Breast Cancer," *Oncology Reports* 30, no. 4 (2013): 1996-2002, https://doi.org/10.3892/or.2013.2660.

CHAPTER 16
OUR GLOBAL IMMUNE SYSTEM
PAUL STAMETS

Mushrooms can have powerful health benefits. Throughout history, indigenous cultures have used mushrooms as medicine, and as the Western world begins to study mushrooms in earnest, we're learning more and more about their incredible potential.

Indigenous to the forests in northwest Washington State is a unique species of mushroom called agarikon. The Latin name is *Laricifomes officinalis*, also known as *Fomitopsis officinalis*. It has been used medicinally for more than two thousand years. In 65 CE Europe, the Greek physician Pedanius Dioscorides, who was also a pharmacologist and botanist, wrote about agarikon in *De Materia Medica*, describing it as the *elixir ad longam vitam*: the elixir of long life. It's on the brink of extinction in Europe and Asia, but I have found specimens in the towering old-growth forests of British Columbia and the US Pacific Northwest.

I've been doing research on agarikon for many years and have the largest strain library of agarikon in the world. Why is this mushroom so interesting to me? Well, it's the longest-living mushroom on the planet. Growing in the depths of old-growth forests, it's subjected to hurricane-force winds and hundreds of inches of rain per year while defending itself against legions of parasitizing bacteria and other fungi. And yet these mushrooms can live up to seventy-five years. How can it survive under so many pressures and yet be able to live longer than many humans? What is the secret to agarikon's longevity?

Working with the National Institutes of Health laboratories, we discovered the reason: a powerful immune system, which revealed a new class of antiviral agents. One of the components of agarikon (and other fungi) that make them so medicinally valuable are polyphenols and other metabolites, which are secreted by the cells, and also happen to be powerful immune system stimulators.[1] Since humans are afflicted by many of the same microbes that harm mushrooms, what we're learning about agarikon's host defenses against pathogens has a lot of applicability. And because medical science long ago discovered that antiviral agents differ in their ability to fend against microbial competitors, having as many agarikon strains available as possible is important. Fortunately, after decades of searching for this elusive species, we have isolated more than sixty strains of agarikon—and we might need every one of them.

OPPOSITE TOP *Lactarius sp.*, Arctic National Wildlife Refuge, Alaska. **OPPOSITE BOTTOM** *Hygrocybe coccinea*, Rachel Carson National Wildlife Refuge, Maine.

🍄 DEFENSE AGAINST THE UNTHINKABLE 🍄

Tuberculosis (TB) is the ninth leading cause of death worldwide, and the rate is growing.[2] There were approximately ten million new cases of the disease in 2016, two-thirds of them in just seven countries: India, Indonesia, China, the Philippines, Pakistan, Nigeria, and South Africa. In 2016, an estimated one million children became ill with TB. Drug-resistant TB is now an increasing problem worldwide, and the treatment success rate is not much better than 50 percent.

One of the reasons Dioscorides sung the praises of agarikon is that it was used to treat respiratory illnesses and "consumption," which we now call tuberculosis.[3] A few years ago, a couple of academic researchers intrigued by this history came to a remote island in British Columbia in search of agarikon. When we found some and examined them in the lab, we found two novel antimicrobial compounds that were specifically active against tuberculosis bacteria. We ended up publishing a groundbreaking paper on our findings.[4]

Years ago, we submitted more than three hundred samples of diverse mushroom extracts to the Bioshield biodefense program administered by the National Institutes of Health and the United States Army Medical Research Institute of Infectious Diseases. Of all the samples submitted, agarikon stood out. Of the seven strains that were tested, a few showed exceptionally strong activity against such viruses as pox (cowpox), swine (H1N1) and bird (H5N1) flu, and herpes (HSV-1, HSV-2). Some of these results were confirmed by a team of Russian researchers, who also found that agarikon is comparatively nontoxic to human cells compared to existing antiviral options.[5] Given our growing resistance to modern antibiotics typically used for a host of conditions, these results are incredibly promising.

ABOVE *Paul Stamets holding a nearly forty-year-old agarikon* (Fomitopsis officinalis *or* Laricifomes officinalis) *from an old-growth forest in Washington.*

Importantly, agarikon could have helped the North American native peoples ward off viral diseases like smallpox and flu viruses that were introduced by European settlers if they had known that the mycelium growing within the tree was excreting these antiviral substances.[6] Although the World Health Organization officially declared naturally occurring cases of smallpox to have ended in 1980, the United States government lists smallpox among the biological agents it most fears terrorists might use. The Food and Drug Administration (FDA) recently approved its first drug intended to treat smallpox—TPOXX, or tecovirimat. Research on tecovirimat began, perhaps not coincidentally, after the 9/11 terrorist attacks.[7]

PROTECTING THE FUTURE

"Incorporating these fungi into our lives and our environment strengthens the defense of the entire ecosystem."

How important are fungi to medicine? The clearest and best example we have is the discovery of penicillin by Alexander Fleming in 1928. He found that when an accidental contaminant got into his cultures of staph bacteria, they inhibited the bacteria as the cultures grew. He wondered what was causing that, which led to his discovery. He received the Nobel Prize in 1945 for his work, but there's a lot more to the story.

The penicillin strains produced in Fleming's and others' laboratories at the time weren't productive enough to make them commercially feasible, so the research continued. Around 1942, a lab assistant in Chicago acquired a moldy piece of fruit from the local grocery and isolated the strain that became the first hyperproducer of penicillin. It turns out that different strains of fungi within the same species can express different amounts of active ingredients, which speaks to the importance of mycodiversity. This all happened during World War II, and the discovery was revolutionary because there were no broad-spectrum antibiotics that could help to heal battle wounds. The Allied powers had these powerful antibacterial strains. The Axis powers did not.

It was such an important discovery, and the scientists were so concerned about losing the strains if their laboratories were bombed that they impregnated the collars of their shirts with spores of the Penicillium mold. It was a well-known fact that wounded soldiers suffered long recovery times (if they recovered at all!) while taking time and resources from medical staff. With this effective new treatment, healing times accelerated, there were a lot fewer deaths, and the strain on precious resources dropped. The Americans and British considered the discovery critically important for national security, since the Japanese and Germans didn't have such a medical tool. In fact, it's been suggested that the discovery of this hyperproducing strain of penicillin was so significant, both medically and economically, that it tilted the balance of the war in favor of the Allied powers.

I believe habitats have immune systems just like people, and that mushroom mycelium is the molecular bridge between the two. Incorporating these fungi into our lives and our environment strengthens the defense of the entire ecosystem, preventing disease vectors from emanating out. We are truly a reflection of the environment in which we grow. If you live in the middle of Shanghai, for example, or in the middle of a polluted industrial site in Pittsburgh, Pennsylvania, you are a

reflection of that environment and you are likely immunocompromised. This makes you a threat to the entire biosphere because you could be the source of the vector of a disease or pandemic that can infect other people.

It is essential that we protect old-growth forests and the biodiversity of their ecosystems because they have within them fungal species that could be critical for human survival. If there's a smallpox pandemic today and our smallpox drugs are inadequate at mitigating the spread, or a virus mutates into a human variant of H5N1 bird flu that sweeps the landscape, you can imagine what might happen. If we lose an old-growth-forest mushroom species like agarikon, which I know is active against those viruses, and it happens to be the one strain—a super strain—that can fight these viruses, what then? Would you rather cut the forest for a few hundred thousand dollars, or save the forest and its resident mushrooms because of their potential to prevent a medical catastrophe?

We occupy this planet together and need to protect what we have because the health of the ecosystem directly affects the health of each of us. We are only scratching the surface of the medicinal potential of mushrooms found throughout the world. Human health and the health of the species that support us are intertwined. So let's keep working collectively to ensure a sustainable future. We really have no other choice.

NOTES

1. Paul Stamets, "Agarikon: Ancient Mushroom for Modern Medicine," HuffingtonPost.com, December 6, 2017.
2. "Global Tuberculosis Report 2017," World Health Organization (Geneva: 2017), http://www.who.int/tb/publications/global_report/en/.
3. Paul Stamets, "Medicinal Polypores of the Forests of North America: Screening for Novel Antiviral Activity," *International Journal of Medicinal Mushrooms* 7, no. 3 (2005): 362.
4. Chang Hwa et al., "Chlorinated Coumarins from the Polypore Mushroom, *Fomitopsis Officinalis*, and Their Activity Against *Mycobacterium Tuberculosis*," *Journal of Natural Products* (2013): 1916-22.
5. T. V. Teplyakov, N. V. Purtseva, T. A. Kosogova, V. A. Khanin, and V. A. Vlassenko, "Antiviral Activity of Polyporoid Mushrooms (Higher Basidiomycetes) from Altai Mountains from Russia," *International Journal of Medicinal Mushrooms* 14, no. 1 (2012): 37-45; Paul Stamets, "Agarikon: Ancient Mushroom for Modern Medicine," HuffingtonPost.com, December 6, 2017.
6. Paul Stamets and C. Dusty Wu Yao, *Mycomedicinals: An Informational Treatise on Mushrooms* (MycoMedia, 2002), 21.
7. Donald G. McNeil Jr., "Drug to Treat Smallpox Approved by F.D.A., a Move Against Bioterrorism," *New York Times*, July 23, 2018.

TOP LEFT Calvatia sculpta (© Taylor Lockwood). **TOP RIGHT** Favolascia pustulosa (© Taylor Lockwood). **OPPOSITE TOP** Tremella mesenterica. **OPPOSITE BOTTOM** Geastrum *sp., commonly known as earthstars.*

SECTION III
FOR THE SPIRIT

We are on a never-ending search for partners . . .
Life-affirming relationships.
Or, at the very least, nourishment for the
next leg of our journey.
We have flourished side by side with your species,
symbiotically, for centuries.

The phenomenon of mystical experiences has generally been associated with religious and spiritual traditions dating back centuries. But for as long as humans have inhabited the planet, there have been compounds in nature that trigger similarly exceptional encounters.

After decades of suppression, research into how mystical experiences influence our thoughts, beliefs, and behaviors is undergoing a renaissance, and researchers in the academic community are using psilocybin and other psychoactive substances to induce and study these remarkable states. The results have been impossible to ignore.

In this section, experts explore the latest research into how mushrooms can help us to understand human consciousness.

THE RENAISSANCE OF PSYCHEDELIC THERAPY • MICHAEL POLLAN (pg.120)

PSILOCYBIN IN A TEST TUBE • NICHOLAS V. COZZI (pg.124)

DOORWAYS TO TRANSCENDENCE • ROLAND GRIFFITHS (pg.128)
SPOTLIGHT ON: TOUCHING THE SACRED • ALEX GREY

A GUIDE THROUGH THE MAZE • MARY COSIMANO (pg.136)
SPOTLIGHT ON: JOHNS HOPKINS SESSION TRANSCRIPTS
SPOTLIGHT ON: SACRED CEREMONY • ADELE GETTY

A GOOD DEATH • STEPHEN ROSS (pg.142)
SPOTLIGHT ON: PSILOCYBIN RESEARCH: A CLINICAL RESURRECTION • CHARLES GROB

THE MYSTERIES OF SELF-NESS • FRANZ VOLLENWEIDER (pg.148)
SPOTLIGHT ON: MICRO-DOSING: REAL OR PLACEBO?

OF APES AND MEN • DENNIS McKENNA (pg.152)
COUNTERPOINT: THE STONED APE THEORY—NOT • ANDREW WEIL

MYSTICAL EXPERIENCES ACROSS RELIGIOUS TRADITIONS • ANTHONY BOSSIS, ROBERT JESSE, AND WILLIAM RICHARDS (pg.156)
SPOTLIGHT ON: COUNCIL ON SPIRITUAL PRACTICES/HEFFTER RESEARCH INSTITUTE
SPOTLIGHT ON: MEDITATION AND PSILOCYBIN • VANJA PALMERS

CHANGING THE GAME • PAUL STAMETS (pg.162)

CHAPTER 17

THE RENAISSANCE OF PSYCHEDELIC THERAPY[1]

༓ ༓ ༓

MICHAEL POLLAN

MICHAEL POLLAN is an author, journalist, activist, and the Lewis K. Chan Arts Lecturer and Professor of Practice of Non-Fiction at Harvard University. He is the author of six *New York Times* best sellers. His latest book is *How to Change Your Mind: What the New Science of Psychedelics Teaches Us about Consciousness, Dying, Addiction, Depression, and Transcendence*.

What we don't know about psychoactive mushrooms far exceeds what we do know. But one thing is becoming certain: Their effectiveness in treating a range of psycho-emotional disorders can no longer be ignored.

While teaching journalism at Berkeley not long ago, I learned about some surprising research involving psilocybin, the active ingredient in so-called magic mushrooms. It was being used to induce mystical experiences as well as treat people who were suffering from end-of-life anxiety and existential distress after learning they'd been diagnosed with terminal cancer.

As I began talking to some of the researchers involved, I learned to my astonishment that the history of this work dates back to the 1950s, and that long before LSD became a drug of the counterculture, it had been embraced in psychiatric circles. Hundreds of peer-reviewed papers had been published detailing its successful use in treating depression, alcoholism, and obsessive-compulsive disorder (OCD), among other conditions. The doctors I was talking with had begun to excavate this buried knowledge. And buried it was. When these psychoactive substances jumped from the lab to the street in the 1960s, unleashing a kind of Dionysian energy and inspiring a growing counterculture to "question authority," the government came down hard, the research community fell in line, and the work essentially stopped.

This was a wake-up call for me. A highly productive area of scientific research had simply been abandoned, which doesn't happen very often. Now, decades later, it's experiencing a renaissance. There's a growing willingness to let people experiment once again with these powerful molecules to see how useful they might turn out to be. And the knowledge of all this inspired me to write a book about it.

When *How to Change Your Mind* first came out, I expected a lot of resistance, but the response surprised me. Our culture seems much more receptive to the idea of psychedelic therapy than it once was; it's no longer seen as a fringe idea. This is in part a measure of the failure of the mental health system to address the growing scale of human suffering from depression, addiction, trauma, and the like, but also because new research is showing the powerfully positive effects that psychoactive experiences are having on people in some very difficult situations.

In fact, the traditionally conservative FDA has been encouraging the research. Psilocybin will still be a controlled substance, but I believe it will eventually be recategorized from a Schedule I drug (highest potential for abuse) to a Schedule IV (Schedule V is the most benign, including, for example, prescription asthma medicines). In short, it's now on a traditional drug-approval path to treat certain conditions.

OPPOSITE *Fractals*.

All of this is happening without the involvement of Big Pharma. If that industry could find a way to make money on it, of course it would, but there are too many reasons it won't or can't. For example, these particular mushrooms are widely grown and relatively easy to find. The molecules involved are in the public domain or in nature, or their patents have expired, and they wouldn't be used chronically; in other words, they aren't drugs people might take every day for the rest of their lives. That's where most of the pharmaceutical industry's research and development money goes: looking for the next antidepressant that is patentable and positioned for long-term use. The treatments being explored for psilocybin and LSD involve one, two, maybe three pills, and that's it. You can't make a lot of money that way. It doesn't look like a very good business model. In general, the pharmaceutical industry likes drugs it can control and that you have to take every day, and that just isn't the case with something like psilocybin.

🍄 PSYCHEDELICS FOR EVERYONE? 🍄

"These psychoactive substances can create experiences for people that change them in fundamentally beneficial ways."

One of the more interesting debates in the psychedelic community right now is whether there's a place for psychedelics for people who aren't suffering from a particular condition. Stated differently, is there a way for psychedelics to be used for the diagnosable betterment of healthy people? These psychoactive substances can create experiences for people that change them in fundamentally beneficial ways. This seldom happens; our personality is pretty much fixed by adulthood, but openness—to new experiences, to new people, to nature—seems to increase after experiencing psilocybin, and the effect endures. In 2006, researchers at Johns Hopkins gave thirty-six volunteers two to three doses of psilocybin at two-month intervals. Sixty percent of them reported having a "complete mystical experience."[2] Fourteen months later, those same volunteers were interviewed, and a similar number reported that the increased sense of well-being they experienced during the initial study had continued.[3]

Not surprisingly, it's very hard to get research dollars to study drugs designed to help people who aren't sick; our drug-development institutions are designed around treating illnesses. But that's not stopping people from taking psychoactive substances, either on their own or with the assistance of what are called "underground guides." I relied on such assistance while writing the book, and the results were quite positive. But the added legitimacy to psychedelic experiences comes with some added cautions, which I address early on in all my talks. Psilocybin is nontoxic and nonaddictive—there are more dangerous drugs in your medicine cabinet—but people do have bad trips (and people at risk for psychoses should never take psychedelics). Taking psilocybin on a guided journey, however, is a fundamentally different experience from the scare stories usually depicted. Many of these guides are highly skilled and use time-tested protocols for safety and comfort. My concern is that if demand for these experiences exceeds the supply of these underground professionals, the potential for things to go wrong goes up.

🍄 THE SUN STILL SHINES 🍄

As a species, I think we're at a crisis point, and whether we can react fast enough to everything that's happening is a big question. I'm not convinced that it's all going to come out fine. But I'm more hopeful than despairing, which may say more about my temperament than represent an accurate reading of the situation. I see a lot of positive things going on, such as more and more people waking up to the importance of biodiversity and planetary awareness. I've seen our ability to design systems that don't just lower our impact on nature but actually heal it. Many people are trapped in a zero-sum view of nature that says either we gain and nature loses, or the other way around. It's a false idea. There are many ways of organizing our relationship with nature that can give us the food we need and the energy we need and at the same time increase biodiversity and soil fertility. The sun still shines, and that's the key—that's nature's free lunch. We just have to figure out a better way to make use of it.

Mushrooms are a big part of that story, but they remain a mystery. In fact, it's amazing what we *don't* know about mushrooms. We know more about bacteria and plants and certainly animals than we do about mushrooms. They are hard to study and haven't received the kind of research attention these other kingdoms have, but they hold great value if we look a little deeper. There's a brilliant chemistry to mushrooms, and endless possibilities. We're just at the beginning of understanding them.

ABOVE Cookeina sulcipes. **OPPOSITE LEFT** Psilocybe zapotecorum. **OPPOSITE MIDDLE** Psilocybe cyanofibrillosa. **OPPOSITE RIGHT** Psilocybe silvatica.

NOTES

1. This essay was adapted from interview transcripts.
2. R. R. Griffiths, W. A. Richards, U. McCann, and R. Jesse, "Psilocybin Can Occasion Mystical-Type Experiences Having Substantial and Sustained Personal Meaning and Spiritual Significance," *Psychopharmacology* (July 11, 2006).
3. J. Lazarou, "Spiritual Effects of Hallucinogens Persist, Johns Hopkins Researchers Report," Hopkins Medicine (July 1, 2008).

CHAPTER 18
PSILOCYBIN IN A TEST TUBE

NICHOLAS V. COZZI

NICHOLAS V. COZZI teaches pharmacology and directs a research program at the University of Wisconsin School of Medicine and Public Health.

Psilocybin alters the way one sees the world, but why? What biochemical processes are involved, and how can we harness this knowledge for healing and transformation?

Psilocybin is a psychedelic compound produced by approximately two hundred species of mushrooms around the world. The fact that psilocybin is found in this many species suggests that it has some kind of basic function in these fungi that might have evolved in response to a common need; psilocybin may play a central role in a biochemical process that is common to all of these fungi.

Psychedelic mushrooms have been used by Mazatec people in central Mexico for spiritual purposes for uncounted centuries. In June 1955, Robert Gordon Wasson and Allan Richardson became the first Westerners to participate in a traditional Mazatec mushroom ritual, led by a local *curandera*, María Sabina. Wasson was an amateur ethnomycologist who also happened to be a vice president at the banking institution J.P. Morgan & Company. Wasson had been studying mushrooms for decades with a particular interest in Russian fungi. As a result of this interest, he and his wife, Valentina Pavlovna Wasson, MD, coauthored the book *Mushrooms, Russia, and History*, which was published in 1957. He also published *Soma: Divine Mushroom of Immortality* in 1972 and *The Wondrous Mushroom: Mycolatry in Mesoamerica* in 1980.

Following his participation in the Mazatec mushroom ceremony in 1955, Wasson described his experiences in an article entitled "Seeking the Magic Mushroom," published in *Life* magazine on May 13, 1957.[1] It was a groundbreaking story as the West's first exposure to the history and purpose of this ancient ritual. Suddenly, millions of people became aware of the use of psychedelic mushrooms to generate mystical consciousness, and thousands of people flooded the area around Oaxaca, Mexico, seeking enlightenment.

On a return trip to Mexico, Wasson was accompanied by French mycologist Roger Heim, who sent samples of the Mazatec mushrooms to Swiss chemist Dr. Albert Hofmann. Dr. Hofmann is the same man who accidentally discovered the psychoactive effects of LSD in 1943 while working with compounds derived from the ergot fungus. In 1958 and 1959, Hofmann and coworkers published scientific papers describing the isolation and characterization of psilocybin and psilocin as the main psychedelic substances in the Mazatec mushrooms and also described synthetic methods to obtain the compounds.[2]

OPPOSITE Psilocybe azurescens.

THE PATH TO LEGITIMACY

"Interest in the therapeutic effects of psilocybin is growing, and it is being investigated as an option for relieving certain psychological ailments, such as anxiety, post-traumatic fear, and addiction."

Pharmacology is the science of drug action. How do drugs do what they do in the body? Where do they act? How long do they last? What are their effects? At the University of Wisconsin Madison, we're conducting a study on the pharmacokinetics of psilocybin and its metabolite psilocin—the time course of their appearance and disappearance in the blood and urine following ingestion by healthy people.[3]

In our study, volunteers take an oral dose of psilocybin in three escalating doses, spaced a minimum of four weeks apart. We collect blood and urine samples at various points during these sessions, and the samples are analyzed by an outside testing laboratory. Because psilocybin itself is very rapidly metabolized and not detectable in blood, the analysis measures the changing levels of psilocin—the active metabolite of psilocybin responsible for its psychedelic effects—of each of the three doses over time. Although we're not looking for specific therapeutic effects, volunteers do participate in pre-experience and post-experience discussions with research clinicians and answer a questionnaire that assesses their subjective experiences during the dosing sessions.[4] Again, in the pharmacokinetic study, we're only interested in how the body handles the psilocybin over time, not in a specific therapeutic benefit.

Psilocybin and its active metabolite psilocin are tryptamines, a chemical family that includes the neurotransmitter serotonin. Serotonin is present in all human beings and in many animals and plants around the world. As a neurotransmitter, serotonin regulates mood, sexual activity, appetite, sleep, memory and learning, body temperature, some social behaviors, and certain cardiovascular processes. Because of its chemical similarity to serotonin, psilocin activates some of the same brain receptors that serotonin acts upon. These receptor effects change the way that neurons function and how they respond to stimuli. This, in turn, leads to altered perceptions and cognition and can sometimes produce mystical or spiritual experiences in people.

Psilocybin mushrooms produce several compounds in addition to psilocybin, including baeocystin and norbaeocystin, which are also known to be psychoactive.[5] They are present in smaller amounts than psilocybin but may affect the mushroom experience by their presence. In our studies, we're using pure psilocybin

ABOVE LEFT *Mushroom species unknown (© Taylor Lockwood).* ABOVE RIGHT *Journeying with magic mushrooms around a campfire is an experience humans have shared from the distant past to the present day.* OPPOSITE LEFT Psilocybe stuntzii. OPPOSITE MIDDLE Psilocybe azurescens, *the most potent psilocybin mushroom in the world.* OPPOSITE RIGHT Psilocybe baeocystis.

synthesized in my laboratory. This raises the question of whether an experience with the psilocybin mushroom is identical to one with pure psilocybin. I suspect that they're not exactly the same.

Why are we doing this study? Interest in the therapeutic effects of psilocybin is growing, and it is being investigated as an option for relieving certain psychological ailments, such as anxiety, post-traumatic fear, and addiction. The work we're doing at UW-Madison is called a Phase 1 study and will assess psilocybin's safety and characterize its pharmacokinetics. This information is required by the Food and Drug Administration as a prerequisite for moving psilocybin into Phase 2 and Phase 3 studies.

These follow-up studies will test psilocybin for efficacy in people with specific mental health issues. A Phase 2 study can involve tens to hundreds of patients, while Phase 3 trials typically involve hundreds or thousands of patients; they are also multicenter studies involving research facilities across the United States. Although there is no hard-and-fast rule that specifies the exact number of test subjects, there must be a large enough patient population to demonstrate safety and efficacy for the medical conditions for which it is intended. The ultimate outcome would be approval by the FDA for the use of psilocybin by qualified clinicians for the specific conditions for which it has been found effective.

IN SEARCH OF THE LUMINOUS

I'm passionate about what I do because I'm interested in consciousness and its relationship to the physical brain. Studying psilocybin and other psychedelic compounds helps me appreciate the relationship between brain chemistry and mystical consciousness, in particular. Because molecules such as psilocybin are well-defined substances and because they alter the way we perceive the world and can sometimes produce transcendental experiences, I believe they are useful tools to gain fundamental insights into how and where the physical, mental, and spiritual come together in human beings. What is this luminous experience we find ourselves in; what is the world?

NOTES

1. R. Gordon Wasson, "Seeking the Magic Mushroom," *LIFE* magazine (June 10, 1957): 100-20.
2. A. Hofmann and F. Troxler, "Identification of Psilocin," *Experientia* 15, no. 3 (1959): 101-2; A. Hofmann et al., "Elucidation of the Structure and the Synthesis of Psilocybin," *Experientia* 14, no. 11 (1958): 397-9; A. Hofmann et al., "Psilocybin, a Psychotropic Substance from the Mexican Mushroom *Psilicybe mexicana Heim*," *Experientia* 14, no. 3 (1958): 107-9.
3. R. T. Brown et al., "Pharmacokinetics of Escalating Doses of Oral Psilocybin in Healthy Adults," *Clinical Pharmacokinetics* 56 (2017): 1543-54.
4. C. R. Nicholas et al., "High Dose Psilocybin Is Associated with Positive Subjective Effects in Healthy Volunteers," *Journal of Psychopharmacology* 32 (2018): 770-8.
5. A. Y. Leung and A. G. Paul, "Baeocystin and Norbaeocystin: New Analogs of Psilocybin from *Psilocybe baeocystis*," *Journal of Pharmaceutical Sciences* 57, no. 10 (October 1968): 1667-71.

CHAPTER 19

DOORWAYS TO TRANSCENDENCE

ROLAND GRIFFITHS

ROLAND GRIFFITHS is professor in the Departments of Psychiatry and Neuroscience at Johns Hopkins School of Medicine, where he has been conducting research on mood-altering drugs for more than forty years.

About twenty years ago, I started a meditation practice that opened a window for me into what some might call spiritual experiences. Both personally and as a scientist, I became very curious to know more about these kinds of experiences, which some people claim can produce enduring and positive transformative effects. I then began reading about the phenomenology of religious experience and mysticism, which led me to reports, mostly from the 1960s, of mystical experiences occasioned by psychedelic drugs such as psilocybin, the principal psychoactive component of the *Psilocybe* genus of mushrooms. The descriptions in those earlier reports were particularly intriguing to me as a research psychopharmacologist with a new interest in spiritual experience.

In 2000, largely out of my personal curiosity about the nature of the spiritual experience, I initiated a rigorous clinical pharmacology study with psilocybin in collaboration with Bill Richards and Bob Jesse. The study compared psilocybin to an active control compound (Ritalin, or methylphenidate) under double-blind conditions. Expectancy bias was minimized by conducting the study in people who were psychedelic naive (that is, without previous experience with psilocybin or similar compounds) and by keeping participants and study staff (including those monitoring sessions) blinded to the conditions under study. While I had a sincere interest in spiritual experiences, as a scientist I was skeptical about what struck me as an overly enthusiastic reporting bias of the earlier generation of psychedelic researchers and proponents.

The results of our initial study and a systematic replication spoke for themselves.[1] The majority of volunteers reported experiences that appeared virtually identical to mystical experiences described by spiritual and religious figures throughout the ages. We used questionnaires developed by experts in the psychology of religion to assess such experiences. There were reports of positive mood, sometimes openheartedness and love, and the transcendence of time and space. These experiences were characterized by a sense of the interconnectedness of all persons and things, sacredness, and an authoritative sense that the experience was more real and true than everyday waking consciousness.

Even months after the sessions, the participants rated their experiences as being among the most personally meaningful and spiritually significant of their lives. Further, participants attributed to these experiences enduring positive changes in moods, attitudes, and behavior, including strikingly prosocial changes. Ratings of participant behavior by friends, family members, and colleagues

OPPOSITE Psilocybe cyanescens.

at work validated the positive enduring effects reported by participants. In my previous research at Johns Hopkins, I had given dozens and dozens of drugs under blinded conditions for decades, but I had never seen anything like this. The phenomenon was fascinating, amenable to prospective scientific study, and seemed to have far-reaching implications for therapeutics, for radical change in life orientation, and for understanding the nature of moral and ethical behavior.

We have continued this line of research in healthy participants in studies examining the psilocybin effects in novice and long-term meditators, and in an ongoing study, in collaboration with New York University, of psilocybin in religious clergy from various faith traditions.[2] Neuroimaging research in collaboration with my Johns Hopkins colleague Frederick Barrett is revealing intriguing insights into the acute and persisting changes in brain function that accompany psilocybin-occasioned transformative effects.

🍄 THERAPEUTIC APPLICATIONS 🍄

"Participants rated their experiences as being among the most personally meaningful and spiritually significant of their lives."

Given that a single, moderately high dose of psilocybin produced such immediate and sustained positive effects in healthy people, we undertook a study to determine if psilocybin would have acute and sustained therapeutic effects in cancer patients who were facing a life-threatening diagnosis and experiencing significant levels of anxiety and depression. The study used a rigorous double-blind, placebo-controlled design in fifty-one patients who received psilocybin. Six months after the final session, 80 percent of the participants continued to show clinically significant decreases in their symptoms of depression and anxiety and significant increases in well-being. Two-thirds reported their experience as one of the top five meaningful, spiritually significant experiences in their lives.[3] We are currently extending this line of research to examine the effects of psilocybin in treatment of major depression. Excitingly, two organizations (the Usona Institute and COMPASS Pathways) have, as of 2018, submitted protocols to the Food and Drug Administration with the goal of developing larger Phase 3 clinical trials, which could ultimately lead to approval of psilocybin as a medicine for treatment of depression. These initiatives will hopefully lead to approval of psilocybin for medical use within the next several years.

With my Johns Hopkins colleagues Matthew Johnson and Albert Garcia-Romeu, I have also been exploring psilocybin as a treatment of cigarette-smoking addiction. We completed a pilot study examining the effects of psilocybin on long-term smokers who wanted to quit but were unable to do so after multiple attempts. Fifteen people received two to three doses of psilocybin in the context of participating in a program of cognitive behavioral therapy for smoking cessation. One year later, ten of the participants were still smoking-abstinent—an unheard-of result in smoking cessation studies.[4] We are continuing to study this important therapeutic indication and we have planned studies to explore possible therapeutic effects of psilocybin in other medical conditions, such as anorexia nervosa and early Alzheimer's disease.

OPPOSITE TOP *CGI rendering of a mycelial web growing through a forest.*

🍄 WE'RE ALL IN THIS TOGETHER 🍄

The finding that psilocybin can occasion, in most people studied, mystical-type experiences virtually identical to those that occur naturally suggests that such experiences are biologically normal. The question arises, why are we wired to have such compelling, seemingly sacred experiences of the interconnectedness of all people and things—experiences that arguably provide the very basis of the ethical and moral codes common to all the world's religions? I think there is something about the mystical experience that relates intimately to the very nature of consciousness itself.

Reflect on the mysterious truth that, if you turn your attention inward, you can become aware that you are aware. When you do so, an indisputable and profound inner knowing arises that is at the core of our humanity: We recognize that we are all in this together, and an impulse for mutual caretaking arises. I believe that exploration of this inner knowing through contemplative and other spiritual practices can result in a profound, uplifting shift in worldview; a waking up to a sense of freedom, peace, joy, and gratitude that many people simply find unimaginable.

Excitingly, psilocybin-occasioned peak experiences appear to provide a model system for prospective and rigorous investigation of these awakening experiences. I am confident that research will reveal underlying biological mechanisms of action and will likely result in an array of novel therapeutic applications. But more importantly, because such experiences appear to be foundational to our ethical and moral understandings, further research may ultimately prove to be crucial to the very survival of our species.

NOTES

1. Roland R. Griffiths, W. A. Richards, U. McCann, and R. Jesse, "Psilocybin Can Occasion Mystical Experiences Having Substantial and Sustained Personal Meaning and Spiritual Significance," *Psychopharmacology* 187 (2006): 268-83, https://doi.org/10.1007/s00213-006-0457-5; Roland R. Griffiths, M. W. Johnson, W. A. Richards, B. D. Richards, U. McCann, and R. Jesse, "Psilocybin Occasioned Mystical-Type Experiences: Immediate and Persisting Dose-Related Effects," *Psychopharmacology* 218, no. 4 (2011): 649-65; Roland R. Griffiths, W. A. Richards, M. W. Johnson, U. McCann, and R. Jesse, "Mystical-Type Experiences Occasioned by Psilocybin Mediate the Attribution of Personal Meaning and Spiritual Significance Fourteen Months Later," *Journal of Psychopharmacology* 22, no. 6 (2008): 621-32.

2. Roland R. Griffiths, M. W. Johnson, W. A. Richards, B. D. Richards, R. Jesse, K. A. MacLean, F. S. Barrett, M. P. Cosimano, and K. A. Klinedinst, "Psilocybin-Occasioned Mystical-Type Experience in Combination with Meditation and Other Spiritual Practices Produces Enduring Positive Changes in Psychological Functioning and in Trait Measures of Prosocial Attitudes and Behaviors," *Journal of Psychopharmacology* 32, no. 1 (2018): 49-69.

3. Roland R. Griffiths, M. W. Johnson, M. A. Carducci, A. Umbricht, W. A. Richards, B. D. Richards, M. P. Cosimano, and M. A. Klinedinst, "Psilocybin Produces Substantial and Sustained Decreases in Depression and Anxiety in Patients with Life-Threatening Cancer: A Randomized Double-Blind Trial," *Journal of Psychopharmacology* 30, no. 12 (2016): 1181–97.

4. M. W. Johnson, A. Garcia-Romeu, M. P. Cosimano, and R. R. Griffiths, "Pilot Study of the 5-HT2AR Agonist Psilocybin in the Treatment of Tobacco Addiction," *Journal of Psychopharmacology* 11 (2014): 983–92; M. W. Johnson, A. Garcia-Romeu, and R. R. Griffiths, "Long-Term Follow-Up of Psilocybin-Facilitated Smoking Cessation," *American Journal of Drug and Alcohol Abuse* 43, no. 1, 55–60.

SPOTLIGHT ON: TOUCHING THE SACRED

ALEX GREY is an artist, author, and teacher whose signature works of transformative art are housed at the Chapel of Sacred Mirrors in Wappinger, New York. He is on the board of advisors for the Center for Cognitive Liberty and Ethics and is the chair of Wisdom University's Sacred Art Department.

OPPOSITE FROM TOP LEFT TO BOTTOM RIGHT Painting, *1998, oil on linen, 30" x 40"*; Oversoul, *1999, oil on linen, 30" x 40"*; Visionary Origin of Language, *1991-98, acrylic on paper, 10" x 14"*; Nature of Mind Panel I, *1995, oil on wood, 8" x 10". (All images © Alex Grey).*

FOLLOWING PAGES *Unidentified mushroom species in colored light.*

Paraphrasing the eminent psychotherapist Stan Grof, psychedelics are to psychology as the telescope is to astronomy or the microscope is to bacteria. To understand our inner worlds, we need the right tools, and psychedelics have been used by both Eastern and Western civilizations for thousands of years to understand the human soul and psyche. I believe they connect us to a kind of divine intelligence and answer questions that materialist science has been ill equipped to deal with.

The British historian Arnold Toynbee devoted his career to studying the rise and fall of civilizations. He believed that a thriving civilization is formed around a spiritual center, and when people wander away from that anchoring identity, the structure begins to break down. He also felt that civilizations evolved and grew when meeting extreme challenges and died when they didn't. "Man achieves civilization," he wrote, "not as a result of superior biological endowment or geographical environment, but as a response to a challenge in a situation of special difficulty which rouses him to make a hitherto unprecedented effort."[1] I would argue that we are now experiencing such a time. Over the centuries, we've lost contact with any sacred understanding of our relationship with the planet. We are not an ecologically intelligent species.

I believe that psychedelics are the best catalysts for awakening this sacred sense of nature and discovering new ways of healing ourselves and the planet, and that the story of psilocybin is one of the most astonishing and promising developments of the twentieth century. Though most of the work has been clandestine until now, its many benefits are beginning to surface, already leading to a cascade of revolutionary insights. Humanity has long had a relationship with these sacred plants, from the mushroom images found in cave art to the ritual use and mushroom stones of the Mayans to the return of mushrooms today as nutrition, medicine, and a doorway to expanded consciousness.[2] There is a visionary mystical reality that we don't understand, and it's trying to help us. I believe that psilocybin is part of a spiritual renaissance that will give us access to the cosmic and divine intelligence of the universe.

Because fungi are one of the earliest forms of life on this planet, they represent a kind of parent to all the complex life forms that followed—including humans. When you ingest psilocybin, you are, in a way, making contact with the history of our evolution as part of the program for awakening consciousness. It's like taking an important vitamin we've been missing. Our visionary capacities have withered, and it's time to re-exercise them and reassure ourselves that we have souls.

—ALEX GREY

NOTES

1. Arnold J. Toynbee, *A Study of History: Volume I: Abridgement of Volumes I-VI* (Oxford U.P.), 570.
2. Brian Akers, "A Cave in Spain Contains the Earliest Known Depictions of Mushrooms," *Mushroom: The Journal of Wild Mushrooming*, accessed August 23, 2018, https://www.mushroomthejournal.com/a-cave-in-spain-contains-the-earliest-known-depictions-of-mushrooms/; F. J. Carod-Artal, "Hallucinogenic Drugs in Pre-Columbian Mesoamerican Cultures," *Neurología* 30, no. 1 (January-February 2015): 42-9; Carl de Borhegyi, "Precolumbian Maya Mushroom Stone Cult," *Pre-Columbian Art*, May 10, 2012, accessed August 23, 2018, https://mayamushroomstone.wordpress.com/2012/05/10/mushroom-stones/.

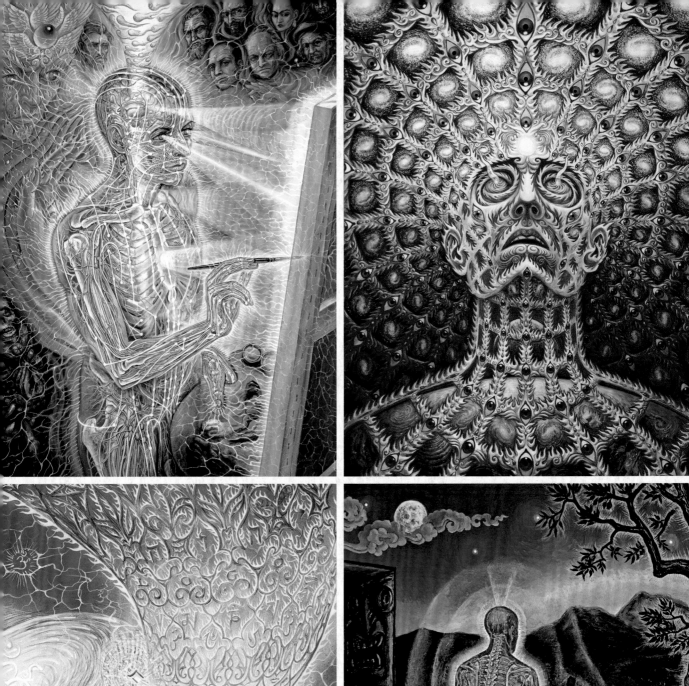

CHAPTER 20
A GUIDE THROUGH THE MAZE

MARY COSIMANO

MARY COSIMANO works in the research and clinical component of Johns Hopkins's psilocybin studies as a guide for participants.

The clinical settings for studies of psilocybin are a critical component of the research process. We essentially create sacred spaces for participants to explore their deepest selves.

When volunteers enroll in one of our psilocybin studies, they're each assigned two guides who accompany them throughout the entire process, from preparation to the actual sessions to the follow-up integration. I've guided about 440 study sessions and one thousand preparatory and integration meetings.

Preparatory sessions typically run eight hours over the course of a few days or weeks, depending on the situation and where a volunteer lives. They focus on developing trust and a rapport with the participant so that he or she will feel safe and open to whatever comes up when we administer the psilocybin. We spend time discussing their life history—their story—which is usually somehow connected to their current condition. Once the psilocybin is ingested in the morning on the day of the session, they spend most of the day lying on a sofa with eye shades and headphones listening to a playlist that is developed specifically for high-dose psilocybin sessions—beautiful, uplifting, flowing music. The music acts as a safe container for them to go inward and stay relaxed and expand into whatever experience the psilocybin brings forth.

It's difficult to explain what I get out of these experiences. In some ways it's as ineffable for me as it is for the participants. At the very least I feel honored just being in such a sacred space. You quickly develop an intimate relationship as these brave volunteers reveal their vulnerability and courage. In bearing witness to their process, I can't help but feel reminded of those very same qualities in me as I help them through challenging moments and share their breakthroughs. In this expansive yet connected space, the heart opens up. Again, it's difficult to put into words. Something primal and beautiful emerges with almost every volunteer we journey with.

OPPOSITE Clavaria *sp*. ABOVE Marasmius *sp*.

SPOTLIGHT ON:
JOHNS HOPKINS SESSION TRANSCRIPTS

JUDITH
A poet and former clinical supervisor at an acupuncture clinic for students; maintained a private practice for twenty years; kidney cancer.

I didn't know that I wanted to subject my sweet, persevering body to this... thing, where there were no promised rewards, but I trusted the team. The change in heart came when I asked the question, "Are the difficulties in these experiences productive? Are they constructive? Or are they gratuitously violent?" When I was told, "Productive," I then asked, "A hundred percent of the time?" The answer: "Yes." I said, "I'm in."

There were images of gothic ceilings, ornate woodwork, rapidly changing colors. I could not access the part of my brain that could describe to the people in the room what was going on. As I struggled to tell them what was happening, they said, "We got it, just be in the experience." I then had a sensation of *becoming* the images; I was no longer observing them, and I had this *yikes!* moment. Is this what it feels like to be possessed? In my mind I said, "Okay, hold it. If I give myself over to you, can you promise me that I will be in at least as strong a shape as when I entered this room?" And when I heard back, "Do you think I would disrespect my own handiwork?" everything lit up. I had an experience of having breath animate me. It felt sacred, almost impossible to describe. The images started changing quickly, and they were fabulous. They felt like they were me. I felt so beautiful, like I've never felt before. I was perfect, but I wasn't flawless...

The visuals continued. There was an image of a tapestry weaving golden threads, every thread needed in the cloth. Every thread there for a purpose. We are all part of this. We're all needed. All of the beginnings and all of the endings, all of the births, all of the deaths, and everything in between are witnessed and loved and cared deeply about. This terrible sense of isolation I'd been feeling for so long dissipated. It was huge.

In my efforts later to make something of the experience, I wrote this small piece called "Prisms":

> *Pure white light passes through living prisms,*
>
> *breaks into radiant, dazzling colors that undulate like neon sea creatures*
>
> *made for the simple pleasure of glowing.*
>
> *We are oddly invisible to ourselves.*

ABOVE *Judith.*

OPPOSITE TOP *Tony.*

OPPOSITE BOTTOM LEFT *Dry psilocybin mushrooms in plastic bag and wooden bowl on brown table.*

OPPOSITE BOTTOM RIGHT *Mycena leaiana.*

TONY

An actor; prostate cancer.

My cancer got outside the prostate and into a lymph node, and when that happens, it's a whole other ball game. The doctors looked at that and basically told me it will probably come back and come back quickly and showed me models and statistics with the odds of how long I might live. Dealing with that was incredibly stressful. I did the best I could but knew I needed help.

Before the session, I was really worried about all my "stuff" coming up, my hang-ups, whatever. Dredging up things that could be disturbing. I was scared at first, and when I felt my arms and legs start disappearing, I thought, "What did I get myself into?"

I don't know what it was exactly—maybe the fact that I've been on this journey for so many years—but I just kept repeating, "Just let yourself go. Just let it go. Release yourself to this." And the more I let go, the more secure I felt, and I started to float along. Then I came in contact with... something, reaching out to me saying, "I'm here." It was incredible. And I had an epiphany: Everything I feared wasn't that important. Connecting with this presence was. I saw that there are laws at work in this world beyond man's laws but also that we can choose to live more positively in a way that relates to people as human beings and not as members of a category. We are all interconnected to each other as well as to the land and to nature.

This experience with a higher power left a mark. It made me feel that there's a purpose for us in this world. I'm now more patient, more willing to think through challenges, to be more understanding. I've tried to live more compassionately and guard against stereotypes and judgments—which culture and society force upon us every day by telling us how to feel and think. Nobody can tell me that my psilocybin experience was *just* a drug. People don't know; it's something you have to experience. It was a direct connection to something marvelous and awesome.

SPOTLIGHT ON: SACRED CEREMONY

ADELE GETTY is a psychologist, author, and ceremonialist. She is the director of the Limina Foundation and supports cognitive liberty and the right to expand consciousness. She believes in the power of psychoactive substances as agents of change.

BELOW *Mixed wild mushrooms.*
OPPOSITE *Fractals.*

Psychoactive plants and molecules have coevolved with human beings. We are made for one another, like a lock and a key. When we take these substances in safe conditions, we unlock a door to perception, and they begin to teach us. They talk to us. They show us. They inform us.

When we use these psychoactive substances in an intentional, sacred way—a ceremonial way—we have the opportunity to move beyond personal psychological challenges and into the visionary or mystical realm where all things become as one. These transcendent realizations—with experiences of unity with the godhead, nature, and self—can both humble us and transform our lives. When these medicines are taken in a natural setting, the experience is often enhanced; it's as if the world of plants, animals, rocks, rivers, and mountains informs us in no uncertain terms that we are a part of the inner-penetrating web of life, a true kinship that is at the core of human experience.

—ADELE GETTY

CHAPTER 21
A GOOD DEATH

STEPHEN ROSS

STEPHEN ROSS is associate professor of psychiatry and child adolescent psychiatry at New York University Langone Medical Center (NYULMC) and director of the NYULMC Addictive Disorders and Experimental Therapeutics Research Laboratory.

For many people, the prospect of facing one's death can be profoundly unsettling, an agony filled with anxiety and despair. But as end-of-life studies of psilocybin have shown, a transcendent experience can change everything.

In medical school, I never received any lectures on death and dying or the concept of "a good death." We're trained instead to help seriously ill people alleviate or at least manage their pain and suffering, but there's no paradigm for helping them with end-of-life distress or the spiritual issues that can come up around that. About 70 percent of Americans die what I would call "bad deaths" in hospitals or intensive care units. The human encounter with death in the United States has been put in the hands of medicine, but doctors aren't adequately trained for it.

In 2006, a colleague started talking to me about ayahuasca, which is a plant-based spiritual medicine used for centuries in sacred ceremonies in the Amazon. As an expert in drugs and addiction, I thought I'd heard of everything, but this was new. This was also the year that the Supreme Court allowed União do Vegetal (UDV), a Christian Spiritist religion that originated in Brazil, to legally use ayahuasca as part of its religious services—similar to the role of peyote in the spiritual rituals of the Native American Church (NAC).[1] And I thought, how interesting that the use of psychedelics for spiritual purposes is protected as religious freedom in the United States, but if you use psychedelics in any other context, they're illegal. What really fascinated me, though, is that within the field of psychiatry, such substances had already been tested for therapeutic purposes. In 1960, for example, Stanislav Grof began using LSD with clients in his psychotherapy practice at the Psychiatric Research Institute in Prague, Czechoslovakia.[2] He also wrote a book about this and subsequent work, *LSD Psychotherapy: The Healing Potential of Psychedelic Medicine*. And so I was inspired to figure out a process and protocol for bringing psychedelics back into a clinical setting and applying them to people who were undergoing the very difficult experience of facing the end of their lives.

> "I was inspired to figure out a process and protocol for bringing psychedelics back into a clinical setting and applying them to people who were undergoing the very difficult experience of facing the end of their lives."

OPPOSITE TOP LEFT TO BOTTOM RIGHT Clathrus archeri; Lysurus periphragmoides (© Taylor Lockwood); Aseroe rubra (© Taylor Lockwood); Geastrum triplex; Deflexula sp. (© Taylor Lockwood); Clathrus ruber; Clathrus ruber; Hydnellum peckii (© Taylor Lockwood).

🍄 ENDURING RESULTS 🍄

We had plenty of patients here at Bellevue Hospital and NYU Langone Medical Center in New York dying these bad deaths, and so my purpose in designing an initial study was to help people have a good death using spiritual interventions to hopefully change the last stages of their lives. We recruited twenty-nine participants for the study, all experiencing advanced degrees of emotional anguish, mental anguish, or both. Their quality of life had become impaired, they had a decreased sense of spiritual well-being, and they'd become disconnected from their sources of spiritual or religious sustenance. Hopelessness and fear of death were common. They were mostly women aged twenty-two to seventy-five who had been doing well in life until receiving their diagnosis (either advanced breast, gastrointestinal, or blood cancer). Each volunteer was provided with counseling.

In the study, we tested single-dose psilocybin against a single-dose placebo, and the results were very promising. A one-time treatment brought dramatic relief in anxiety, depression, and orientation toward death that lasted to the end of the study's eight-month monitoring period for 80 percent of the participants. Other reported improvements included being more outgoing, having greater energy, getting along better with family members, and doing well at work. Sustained feelings of spirituality, peacefulness, and altruism were also reported. The overall project from start to publication spanned a decade.[3]

☙ MYSTICAL HEALING ☙

Because we've been able to show statistically that the intensity of the mystical experience is highly correlated to clinical improvement, we're trying to understand the basic mechanisms behind it. What specific aspects of a mystical experience decrease cancer-related anxiety and depression? What phenomenology is driving these therapeutic effects?

We all have this illusion that we'll live forever, that death is "over there." A cancer diagnosis shatters that, provoking a kind of, "Oh my god, I'm going to die. What's my life about?" response. This fracture of the spiritual domain is similar to what happens in addiction, where the impact of spiritual deficits comes up often. That's why I thought the fit would be good between people experiencing the trauma of dying and death and the potential of psilocybin to open them up to other perspectives. But what, exactly, is the mechanism of change?

When someone is facing death, they experience a sort of dissonance between their ego—consciousness as finite and contained—and something entirely foreign, another kind of consciousness. The psilocybin experience tends to dissolve the ego and the boundary between self and others. It gives people a profound sense of oneness, of their consciousness as part of something greater. In our cancer study, patients would report that they were no longer scared of dying, that they saw consciousness as continuous, as part of a bigger reality. And so perhaps that experience of oneness in the mystical state is the key component that helps a dying cancer patient overcome distress.

Another possibility is that people will have these profound intuitive insights, a strong sense of, "Oh, so that's what's really going on," or, "I get it now." Epiphanies of deep personal truth. For a lot of reasons, psilocybin puts people in touch with things that are sacred and meaningful and genuine, with authentic experiences of love that have powerful healing effects.

ABOVE *Gathering around a campfire for a ceremony.* **OPPOSITE** *The liberty cap,* Psilocybe semilanceata.

GROWTH INDUSTRY: PSYCHEDELIC THERAPY

A recent population-based survey of several hundred thousand people looked at the relationship between drug use and suicidal behavior.[4] It found that every drug of abuse such as alcohol, tobacco, and cocaine was associated with greater rates of suicidal thinking or behaviors—except psilocybin, MDMA, and other serotonergic hallucinogens. Those were associated with *lower* rates of suicidal thinking and behaviors. Does this mean I think everyone should be taking these substances?

You have to be very careful when you start introducing psychedelics into the population. We got into trouble with that in the '60s. It moved too quickly, wasn't done in the right way, and caused unnecessary harm. And yet there's nothing specific about cancer that makes it unique to anxiety around death. Anyone with a serious medical illness could be similarly affected, and of course you don't need to be sick to have existential distress around death. Freud thought that a core fear of human beings was fear of death. And so to help with that, for spiritual growth, these

substances can be applied in the general population, but it would have to be done carefully, there would have to be some screening, and it would have to take place in a specific kind of clinical setting.

I anticipate a new industry within the mental health field: training to become a psychedelic therapist. Psychedelic research has been resurrected and validated with the help of a small group of people who kept the flame alive. We now have real data from rigorous studies that are impossible to ignore. Prestigious academic and medical institutions—including Johns Hopkins, NYU, UCLA, the University of New Mexico, the University of Alabama, Yale, UCSF, the University of Arizona, and the University of Wisconsin—keep driving this research forward. These compounds can be enormously beneficial, and we're just now beginning to truly understand how and why. And because it's happening in mainstream academia, there is no turning back.

NOTES

1. The official website of União do Vegetal (UDV), http://udvusa.org/.
2. Alexander Zaitchik, "How Stanislav Grof Helped Launch the Dawn of a New Psychedelic Research Era," AlterNet.org, April 9, 2010.
3. Stephen Ross et al., "Rapid and Sustained Symptom Reduction Following Psilocybin Treatment for Anxiety and Depression in Patients with Life-Threatening Cancer: A Randomized Controlled Trial," *Journal of Psychopharmacology* 30, no. 12 (2016): 1165-80; "Single Dose of Hallucinogenic Drug Psilocybin Relieves Anxiety and Depression in Patients with Advanced Cancer," press release, NYU Langone Medical Center, December 2016.
4. "Suicidal Thoughts and Behavior among Adults: Results from the 2014 National Survey on Drug Use and Health," Substance Abuse and Mental Health Services Administration, September 2015.

CHARLES GROB is a professor of psychiatry and biobehavioral sciences and pediatrics and director of the Division of Child and Adolescent Psychiatry at Harbor–UCLA Medical Center.

TOP LEFT *Psilocybe cubensis growing in the wild.*
TOP RIGHT *Psilocybe cyclonic.*
OPPOSITE *Calvatia craniiformis.*

 SPOTLIGHT ON:
PSILOCYBIN RESEARCH: A CLINICAL RESURRECTION

From 2004 to 2008, my colleagues and I at the Harbor–UCLA Medical Center carried out the first psychedelic treatment study of advanced cancer anxiety in nearly forty years.[1] Because this kind of research had been shut down for so long, it was incumbent on us to demonstrate feasibility and prove that this kind of research was not only doable, but also showed promise. Otherwise, future studies would be in jeopardy. Fortunately, we were able to obtain all the necessary approvals and funding, conduct the study, collect and analyze the data, and publish in the professional literature. Also—and very importantly—we established strong safety parameters, and none of the subjects had an adverse physiological or psychological response to the treatment.

The study itself was a pilot project utilizing a double-blind methodology to test the effects of an experience with psilocybin on twelve people with a terminal diagnosis who were suffering from severe anxiety. The results were very promising. We found a statistically significant decline in anxiety that lasted for months, as well as a significant correlation with improved mood that was evident for many months after the treatment session. We also found improvements in other measures of mental and emotional health, including psychological status, diminished demoralization, and improved quality of life, in their remaining time. Essentially, we met all the goals we set for the study that replicated many of the findings of the pioneers in this research going back to the 1950s and '60s.[2]

—CHARLES GROB

NOTES

1. C. S. Grob et al., "Pilot Study of Psilocybin Treatment for Anxiety in Patients with Advanced-Stage Cancer," *Arch Gen Psychiatry* 68, no. 1 (January 2011): 71-8.
2. C. S. Grob, "Commentary on Harbor–UCLA Psilocybin Study," *MAPS Bulletin* 20, no. 1.

CHAPTER 22
THE MYSTERIES OF SELF-NESS

FRANZ VOLLENWEIDER

FRANZ VOLLENWEIDER is the codirector of the Center for Psychiatric Research, director of the Neuropsychopharmacology and Brain Imaging Unit, and professor of psychiatry in the School of Medicine, University of Zurich. He is also the director of the Heffter Research Center Zürich, which he founded in 1998.

Are mystical experiences real or a product of physical processes? Who or what is having the experience? The quest to understand what makes us human has enduring philosophical import, and psychedelic research is leading the way.

Is there a common denominator across different drugs and inducing agents that produce what we call an "altered state of consciousness"? Does it depend on the brain, one's environment, or the culture? Or is there some innate phenomenon that correlates with a specific kind of drug? During the last twenty years or so, my colleagues and I have used several different approaches, from brainwave studies and PET (positron-emission tomography) scans to fMRI, in trying to understand the neural basis of altered states. So far, we have found that different qualities common to most experiences can be linked to a specific pattern of brain activity.

When you consider mystical types of experiences, the scientific question is whether they are real or the product of some mix of physical and environmental properties. What role does one's culture play, for example? Are language skills required? What is the influence of genetics and specific environments? Is it primarily a constructed experience based on one's personal associations and history, or an inner experience characterized by more universally shared qualities? Delving further, can we use the power of these drugs to induce specific states? To begin to answer at least some of these questions, we took a closer look at the kind of neuroreceptors that are assimilated by psilocybin.

Psilocybin interacts closely with the serotonin receptor system which, among other things, regulates mood.[1] We've studied about eight hundred mostly healthy people over the years using a range of doses, and our research has found that psilocybin shifts "emotional biases," which is very important in the treatment of depression. When patients have what is called a negative bias, their internal thinking gets stuck in ruminating loops of negative thoughts. We've shown that psilocybin can break that loop down, shifting emotional patterns from negative to positive, suggesting an antidepressive potential that could eliminate the need for traditional antidepressants.[2]

In all our cases we've never had a "bad trip." About one in twelve have a more difficult time than others, but we always guide them through the negative aspects of the experience. In the end, even the challenging parts are integrated as part of the learning and growing process during the follow-up work we do with participants the next day.

OPPOSITE Hygrocybe conica.

In our most recent study, most participants continued to report positive changes in their relationships and a deeper understanding of altered states when queried up to a year later. About 40 percent described themselves as more open and more communicative with friends, and more appreciative of nature and the arts. Specific descriptions included feeling more "integrated" with nature, slowing down and being more aware of what's happening in the moment, developing a growing sense of mindfulness, and feeling more empathetic.[3] It's really impressive that such changes can occur from a single experience.

We have also been interested in how the brain is involved in creating a sense of self. Because psilocybin reduces self-other boundaries, it raises such questions as, "How are we constructing our world? How are we making a self that relates both internally and externally?" This is an especially important consideration for treating patients who suffer from disturbed-self disorders such as schizophrenia. Our psilocybin research is giving us more tools for understanding the brain and psychology and for testing new ideas on how a particular drug or antipsychotic might work. These substances are relatively easy to study in precise experimental situations and have enormous power for helping us investigate the mysterious process of self-ness. This leads more and more into the philosophical aspects of the experience.

And so in addition to continuing to investigate psilocybin's potential for treating a range of mood and anxiety disorders, I think the future of psychedelic research will also explore such complex questions as whether the experience itself is real or a phenomenon of the brain constructing reality and what constitutes the self, and how that may be different from what we understand as the ego.[4] This research can take the investigative process far deeper into questions about human identity and existence.

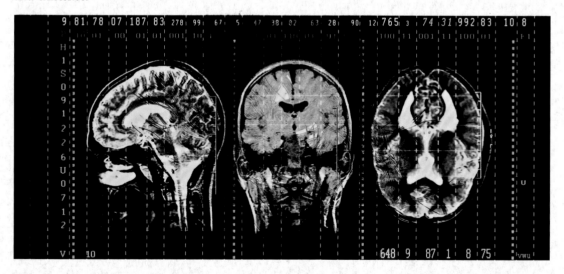

NOTES

1. Rainer Kraehenmann et al., "The Mixed Serotonin Receptor Agonist Psilocybin Reduces Threat-Induced Modulation of Amygdala Connectivity," *NeuroImage: Clinical* 11 (2016): 53-60.
2. Ana Sandoiu, "Magic Mushrooms: Treating Depression without Dulling Emotions," *Medical New Today* online, January 16, 2018.
3. F. X. Vollenweider et al., "Effect of Psilocybin on Empathy and Moral Decision-Making," *International Journal of Neuropsychopharmacology* 20, no. 9 (September 1, 2017): 747-57, https://doi.org/10.1093/ijnp/pyx047.
4. Ananya Mahapatra and Rishi Gupta, "Role of Psilocybin in the Treatment of Depression," *Therapeutic Advances in Psychopharmacology* 7 (2017): 54-6.

SPOTLIGHT ON:
MICRODOSING: REAL OR PLACEBO?

It's no secret that Steve Jobs, founder of Apple, used psychedelics, including psilocybin, as a creative tool, once claiming that "LSD was one of the two or three most important things I have done in my life." Now the use of very small doses of psychoactive substances has become a legitimate workplace tool, and not just in Silicon Valley. The pioneering psychologist James Fadiman has used the term "performance psychedelics." Others prefer "psychovitamin" or "bio-hacking." But does it really work?

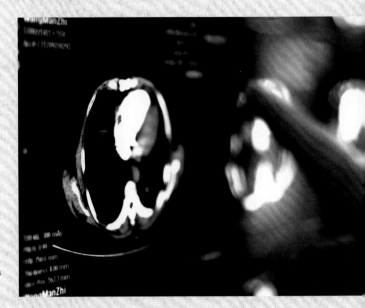

According to Don Lattin, author of *Changing Our Minds: Psychedelic Sacraments and the New Psychotherapy* (2017), "Microdosing is the new thing. Some people will microdose with mushrooms, some with LSD, some with mescaline (from the peyote cactus). It's basically taking a tenth of a dose, and the effect is supposed to be subliminal. If you feel high at all, you've taken too much. There's been plenty of anecdotal evidence that it can improve cognitive function and creativity, but unfortunately there is no good research on it.[1] How often has someone just convinced themselves that they're having an unusually good day?"

William Richards, a clinical psychologist at Johns Hopkins, agrees: "My gut feeling is that it might be helpful for well-integrated people for short periods of time but detrimental to others. We know with moderate-to-high dosages that nondrug variables such as a person's psychological health or the quality of their interpersonal relationships are of critical importance in terms of having beneficial experiences. I don't see why that wouldn't be true of subthreshold doses. Then there's the placebo effect, the power of suggestion. And how do you measure creativity? These are complex questions that need to be investigated in a well-designed double-blind study."

Those questions may soon be answered, at least for LSD. The Beckley Foundation in the United Kingdom is collaborating with leading scientists and universities in the United States, the Netherlands, Brazil, and the UK to develop a research program for investigating the therapeutic potential of LSD at a range of doses. Beckley founder and lead investigator Amanda Feilding believes that by initiating dose-response studies, "We will help determine which doses are likely to produce the best results for different conditions and therapies."

ABOVE *A doctor analyzes an Xray image of the brain.*

OPPOSITE *MRI scans showing the left, top, and front of the brain.*

NOTE

1. Ayelet Waldman. *A Really Good Day: How Microdosing Made a Mega Difference in My Mood, My Marriage, and My Life* (New York: Knopf, 2017).

CHAPTER 23
OF APES AND MEN

DENNIS McKENNA

DENNIS McKENNA is a pharmacologist and a teacher at the University of Minnesota in the Center for Spirituality and Healing. He's been affiliated with the Heffter Research Institute since it was founded in 1993.

The human brain tripled in size over a relatively short evolutionary timespan. Perhaps the "altered-state" inducements of ancient psychoactive mushrooms had something to do with that.

My late brother Terence and I first encountered psilocybin mushrooms in 1971, when we traveled to South America. We brought spores back, learned how to grow them, and a couple of years later wrote *Psilocybin: Magic Mushroom Growers Guide*. We wanted people to have the same experiences we did and confirm that we were onto something. They did, and we were. It turned out to be quite an influential book. And yet psilocybin arrived with much less fanfare and hysteria than LSD, which kind of dropped onto society like a bomb. Mushrooms showed up quietly, as they tend to do. They were the perfect psychedelic after the drama of LSD. They are nontoxic and a lot less dangerous, and most people have enjoyable experiences—with proper doses. They represented a relatively benign way to educate society about what these substances do. But I don't take them lightly. I have great respect for mushrooms. Some people assume that mushrooms are "recreational" and not a risk. This is a dangerous mistake. They are serious and powerful facilitators of altered states of perception.

ABOVE *Unidentified mushroom species.* OPPOSITE TOP LEFT TO BOTTOM RIGHT Psilocybe azurescens; *Fly agaric* (Amanita muscaria); *Fly agaric* (Amanita muscaria); Psilocybe azurescens.

🍄 THE STONED APE HYPOTHESIS 🍄

The stoned ape hypothesis is hard to discuss because there are many layers of complexity to it. And it's an unfair characterization because people interpret it to indicate that early hominids ate mushrooms and suddenly became smart and started talking. It's not that simple, but I do think it's clear that hominids and nearly every creature in the biosphere evolves within a chemical ecology that features a vast array of molecules and microbes, some of which impact the nervous system.

Hominids were scavengers, and mushrooms would have been easy to spot since any grazing animal could provide a potential substrate for the psychoactive varieties through their feces. You simply bend over, pick them up, and eat them. If you picked up a *Psilocybe cubensis*—a fairly common species—you would have had these profound experiences. Given that concepts of a deity and a relationship with nature are built into the indigenous worldview, imagine the impact such an experience must have had on early hominids, however rudimentary their notions of the divine.

One thing that mushrooms and other psychedelics reliably do is induce synesthesia—the conflated perception of one sensory modality with another, for example "hearing colors" or "seeing music." This is important because sounds and colors are inherently meaningless, but when you combine them, a symbol is created. Psychedelics were the trigger in associating sounds and sight to meaningful symbolic complexes. Another of the characteristics of psilocybin—or any psychedelic experience—is portentousness, an experience of momentous significance, even about ordinary things. Psychedelics cause people to notice things they don't normally see. They bring the background forward.

And so it might have been for these early hominids. Over the course of several million years, the human brain tripled in size from approximately 500 cubic centimeters (cc) in the earliest prehumans to 1,500 cc today.[1] This represents a very short span of evolutionary time. What triggered that? It's impossible to prove or disprove, of course, so we're free to speculate, but I think psilocybin was a tool for learning cognition. It was essentially the software that programmed the neurological hardware to think, to have cognition, to have language, and language is essentially synesthesia, the association of an inherently meaningless uttered sound with a complex of meaning.

ABOVE *Rendering of early hominids finding a mushroom.* OPPOSITE *Rendering of a psychedelic experience.*

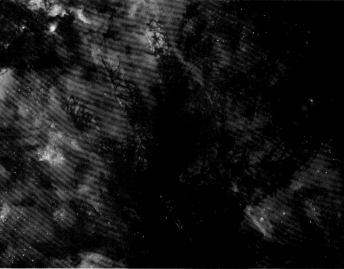

Again, it's not so simple as their brains suddenly mutating by eating psilocybin mushrooms, but I think psilocybin was a factor on the epigenetic level, influencing the development of this neural hardware. A neuroanatomist will tell you that a great deal of the brain's real estate is devoted to the generation and comprehension of language; such neural structures aren't found in our primitive ancestors. This is a modern human trait and a reflection of very recent evolutionary advancement. I believe that mushrooms—psilocybin—could have been a catalytic influence in that.

NOTE

1. Andrew Du et al., "Pattern and Process in Hominin Brain Size Evolution Are Scale-Dependent," *Proceedings of the Royal Society*, no. 1873 (February 28, 2018), 285, https:// doi.org/10.1098/rspb.2017.2738.

COUNTERPOINT: THE STONED APE THEORY—NOT

I don't buy the "stoned ape" hypothesis and never have. The perceptual distortions and sensory scrambling that psychedelics cause would be a fatal risk in a developing species, making you more vulnerable to being eaten by a predator or dying accidentally. I think human beings had to wait until there was some minimal level of civilization and protection in which you had the luxury of space and time to safely explore such experiences.

Human beings are capable of experiencing and appreciating the mystery and the wonder of the divine without substances. You look up at the night sky or observe birth and death, and you're starkly confronted with the mystery in which we live. I think there are all sorts of reasons throughout history as to why the brain has increased in size. Even before there were primates, the brain was developing and growing without any dependence on psychedelic substances.

—ANDREW WEIL

CHAPTER 24
MYSTICAL EXPERIENCES ACROSS RELIGIOUS TRADITIONS

ANTHONY BOSSIS, ROBERT JESSE, AND WILLIAM RICHARDS

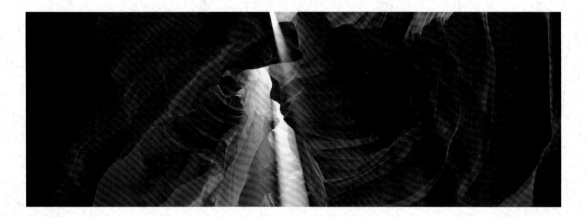

Are mystical experiences consistent across religious traditions? And how might such an encounter influence the work of ordained practitioners, no matter their training and background?

Supported by the Council on Spiritual Practices (CSP), researchers at Johns Hopkins University and New York University are conducting a Phase 1 study to explore the effects of psilocybin on religious leaders to understand how they interpret a mystical-type experience and whether it may enrich their work as clergy.

The two study sites are recruiting twenty-four leaders from a variety of spiritual traditions. After careful screening and preparation, each participant receives psilocybin during two individual, daylong contemplative sessions in the company of trained staff in a comfortable, living room–like setting.

Researchers hypothesize that because of their interests, training, and practices, these religious leaders will be able to offer unique insights, which will contribute to the scientific understanding of both the mystical-type experiences and their impacts on those who go through them.

The study's lead principal investigator is Roland Griffiths, professor in the Departments of Psychiatry and Neurosciences at the Johns Hopkins University School of Medicine. The study's senior guide at NYU is Anthony Bossis, a clinical psychologist. The study's senior guide at Hopkins is William Richards, a clinical psychologist. At NYU, the site principal investigator is Stephen Ross, associate professor of psychiatry and child and adolescent psychiatry. Robert Jesse is convenor of CSP, the study's sponsor.

ABOVE *Sunlight shining through canyons.* OPPOSITE FROM TOP LEFT TO BOTTOM RIGHT *Unidentified mushroom species; Parrot toadstool (*Gliophorus psittacinus*); Marasmiellus sp. on berry cane; Unidentified mushroom species.*

"I'm struck by the fact that human beings seem wired for the kind of incredible, meaning-making, transcendent experiences that are at the core of the world's major religions. Psilocybin, when competently administered with attention to mental set, environmental setting, and appropriate dosage, has been found to reliably facilitate such mystical experiences. Because many have said that institutionalized religions have drifted from their mystical core, my colleagues and I had a radical thought: What would we learn if we invited a diverse group of spiritual leaders—ordained practitioners from Christianity, Buddhism, Judaism, Hinduism, and other traditions—to go through this important doorway? Most of them have spent the better part of their careers studying this landscape. Would there be a common experience? If so, what? If not, how would they be different? What might we discover about the genesis of human spirituality?"

—ANTHONY BOSSIS

"In conceiving this study, we thought it might be useful for religious leaders to have an actual spiritual experience. There is certainly a long history of such intimate revelations across religious boundaries if you can imagine Moses's burning bush, Isaiah's temple visions, and St. Paul's vision on the road to Damascus. And many mystics throughout the ages, such as Teresa of Avila, Meister Eckhart, Hildegard von Bingen, Plotinus, and Rumi, had direct experiences with some transcendent energy, something much bigger than them, and the impacts were profound. Perhaps a pastor would preach with more confidence on Sunday morning or a minister would bring greater depth to their work with the dying. In various ways, spiritual practitioners might appreciate more deeply the richness of their own tradition while being more tolerant of the spiritual paths of others."

—WILLIAM RICHARDS

"Psilocybin, used carefully, is proving to be a wonderful tool for prospectively investigating the nature and consequences of mystical-type experiences. In recent years, researchers in this area have made great strides. In light of increasingly robust scientific findings, surely the culture will come to accommodate these modalities where they prove useful in psychiatry and, when appropriately safeguarded and supported, for the betterment of well people."

—ROBERT JESSE

SPOTLIGHT ON:

COUNCIL ON SPIRITUAL PRACTICES

The mission of the Council on Spiritual Practices is to identify and develop safe and effective approaches for eliciting "religious experiences" that will help individuals and spiritual communities "bring the insights, grace, and joy that arise from direct perception of the divine into their daily lives." CSP has no doctrine or liturgy of its own and is not affiliated with any spiritual or religious tradition. It was first was convened in 1993 by former technology executive and now psychedelic investigator Robert Jesse.

ABOVE Coprinellus disseminatus.

OPPOSITE LEFT Mycena chlorophos.

OPPOSITE RIGHT Purple mushrooms at varying stages of maturity, species unknown.

HEFFTER RESEARCH INSTITUTE

The Heffter Research Institute was incorporated in New Mexico in 1993 as a nonprofit scientific organization. Since its inception, Heffter has been helping to design, review, and fund leading studies on psilocybin at prominent research institutions in the United States and Europe. That research has focused primarily on the application of psilocybin for the treatment of cancer-related distress and addiction and for basic research into the physiology of brain activity, cognition, and behavior. The institute believes that psychedelics have great, unexplored potential that requires independently funded scientific research to find the best uses in medical treatment.

SPOTLIGHT ON: MEDITATION AND PSILOCYBIN

Psychedelics were crucial in my development as a young man. In fact, they completely changed my life. I dropped out of university, stopped the business I was running, and went on a search. That landed me first in yoga and then in Zen meditation. I became a Zen priest, stopped taking psychedelics for about three decades, then recently rediscovered them when I participated in a study in 2015 in which half of a group of longtime meditators were given a dose of psilocybin after four days of sitting meditation practice; the other half were given a placebo.[1] Some of the very experienced people said it was interesting but not fundamentally different from what they'd experienced in their practice; others with similarly long histories of meditation were completely blown away. They knew what was going on but had released control of their bodies and had ecstatic experiences.

I think the combination of a traditional meditation practice with the use of psychedelics holds a lot of promise for the positive development of consciousness. The study suggested that people who have such training are more relaxed about unusual states of mind. If you're on a psychedelic journey with a practiced experience of sitting still and letting things pass through you, of coming back to your breath, of keeping an open presence and a willingness to meet whatever comes up, that can take you to some very beautiful places. Some of the study participants later told me that for a long time afterwards, their meditation practice was different. Their psilocybin experience somehow renewed it and made it more lively again.

These experiences are not a guarantee of long-lasting change, however. The crucial part is what you do with them. This is true if your spiritual breakthrough comes from taking a substance or from sitting quietly for a week without one. The challenge is how to integrate that experience into your daily life. Perhaps a combination of the two helps to anchor those learnings. Another study added two psilocybin sessions to a training program for first-time meditators and found that the two were powerfully complementary.[2]

I believe that mushrooms are like our older planetary brothers and sisters. As a species we are very young, barely in puberty, and behaving accordingly. But when we approach these elders respectfully and carefully, they will talk to us, invite us into their world, and give us instructions. And mostly, what they seem to be saying is to take care of our environment and this planet.

—VANJA PALMERS

VANJA PALMERS is a retired Sōtō Zen priest, dharma heir to the late Kobun Chino Otogawa, and longtime animal rights activist. Along with David Steindl-Rast, a Catholic Benedictine monk, he cofounded Ökumenisches Haus der Stille (House of Silence) at Puregg in the Austrian province of Salzburg, a place of interreligious dialogue between Christians and Buddhists. He is also the founder of Felsentor.

ABOVE *Buddhist prayer wheels.*
OPPOSITE TOP LEFT *Buddhist monk praying.*
OPPOSITE TOP RIGHT *Unidentified mushroom species.*
OPPOSITE BOTTOM *Calocera viscosa.*

NOTES

1. Milan Scheidegger et al., "Psilocybin Enhances Mindfulness-Related Capabilities in a Meditation Retreat Setting: A Double-Blind Placebo-Controlled fMRI Study," Presentation at Psychedelic Science 2017, http://psychedelicscience.org/conference/interdisciplinary psilocybin-enhances-mindfulness-related-capabilities-in-a-meditation-retreat-setting-a-double-blind-placebo-controlled-fmri-study.
2. Roland R. Griffiths et al., "Psilocybin-Occasioned Mystical-Type Experience in Combination with Meditation and Other Spiritual Practices Produces Enduring Positive Changes in Psychological Functioning and in Trait Measures of Prosocial Attitudes and Behaviors," *Journal of Psychopharmacology* 32, no. 1 (January 1, 2018): 49-69.

CHAPTER 25
CHANGING THE GAME

PAUL STAMETS

Mushrooms' mysterious capacity to elicit transformational states of awareness has revolutionary potential. The future of planetary sustainability may depend on the convergence of human consciousness with mycelial intelligence.

I personally believe that psilocybin mushrooms provide medicine for the soul and for the psyche. They help you come out on the other side of issues that may have been dragging you down, that may have harmed you and others. A lot of that healing has to do with forgiveness and moving forward. We can't get stuck in the past; we need to be grounded in the present but not stop looking ahead. If we can visualize a better future, we can achieve a better future.

One of the things that psilocybin does is quiet the brain. As David Nutt, a professor of neuropsychopharmacology at Imperial College London once said, "Psilocybin does in thirty seconds what antidepressants take three to four weeks to do."[1] Rather than having to manage the mass of signals that flow back and forth between the typically more active regions of the brain, the brain partially powers down during a psilocybin experience, allowing other regions to communicate with each other. So, in a way, the altered states that occur are the result both of mind expansion and mind contraction: Turn down the dial on some of the noise so that other signals can be heard. What then tends to arise, as the research has shown, is an experience of oneness, of connectedness with nature. Mushrooms give us access to a spiritual universe, to a conception of God, that we would otherwise never experience. That's usually a game changer.

"We're on the cusp of a mycological revolution that will have paradigm-shifting effects well into the future."

Findings from clinical studies on the benefits of psilocybin for individuals with terminal cancer, end-of-life anxieties, post-traumatic stress disorder, obsessive compulsive disorders, and other difficult conditions continue to accumulate. These are double-blind, scientifically rigorous protocols that are generating statistically significant results. From a therapeutic perspective, psilocybin and other psychoactive substances are doing in four to six hours what would otherwise take years with more conventional treatments (if they are effective at all), and some folks don't have that much time. And I think this is just the tip of the iceberg.

OPPOSITE Coprinellus disseminatus *group.*

A PERSONAL JOURNEY

My brother once returned from Mexico with incredible stories about magic mushrooms and religious experiences. He knew I was into the subject but was a little concerned because I was overly eager. Finally, in Ohio, I purchased a bag.

I had no guide and no recommendations for how much to take, so I ate the entire thing while walking in some nearby woods I knew pretty well. I found out later that I took ten times the "normal" dosage.

It was a warm summer day, and I noticed some black clouds on the horizon. There was this big, old, beautiful oak tree on top of a ridge where I'd spent a lot of time. I really loved that tree. It had a beautiful view of the rolling hills and the pastoral setting of rural Ohio. And I thought, "I'm going to climb to the top of this tree. It will give me a great view of this storm."

I started to feel the effects then; waves of increasing intoxication rolling through me. I climbed to the very top of this tree and watched those boiling clouds come closer, but now they were looking angry. The lightning strikes began, and I started seeing all these geometrical fractals emanate out from them. It was incredible; of course, I'd never seen anything like it.

The wind increased, the rain began, and the lightning was generally getting more intense, and I got scared. I'm on the top of a tree, on the top of a hill, in the middle of a lightning storm. I started getting vertigo and held on to the tree for dear life.

Up to this point I'd had a terrible stuttering habit. I couldn't speak a single sentence without stuttering profusely. I went through six years of speech therapy, but nothing helped. So I'm stuck on top of this tree, and I thought to myself, "What should I focus on? What am I getting out of this experience? Wait! I need to stop stuttering. That's my problem." I couldn't date, I wasn't confident, I didn't feel comfortable in public. And a small inner voice in my head said, "Can you hear me? Stop stuttering now." And I started repeating it—"Stop stuttering now. Stop stuttering now."—hundreds, thousands of times.

Hours later, I crawled down from the tree after the storm had finally passed, soaked to the bone and in love with life, with nature, with that tree. I went home and went to bed.

There was this attractive woman I knew and liked a lot, but I could never look her in the eyes because I was afraid to stutter and embarrass myself. She liked me as well, but I didn't know what to do with it, with that attention. Better just to avoid social contact. So she walked past me that morning, looked at me, and said, "Good morning, Paul." And for the first time ever, I looked her straight in the eye and said, "Good morning. How are you?" I'll never forget that moment. I had stopped stuttering in one session. It still comes back occasionally, but it was 99 percent cured in one six-hour session, although six years of speech therapy couldn't touch it.

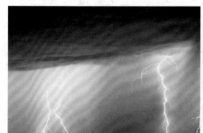

ABOVE LEFT *Lightning storm.* **ABOVE MIDDLE** *Rendering of lighting and trees.* **ABOVE RIGHT** *Lightning in front of a fractal.* **OPPOSITE** *Rendering of the underground mycelial network.*

FUNGAL INTELLIGENCE

We're on the cusp of a mycological revolution that will have paradigm-shifting effects well into the future. There's a concept in science that's very popular right now called epigenetics, which describes how changes in an organism can be caused not by changes in their genetic code but by changes in the *expression* of those codes, and that those changed expressions can be caused by their environment. Fungi's incredible ability to heal environments that have suffered from catastrophic change might be explained by the epigenetic qualities of their network-like design.

Even small clumps of mycelium can have trillions of branching ends on their outer periphery. As these branches grow out into an ecosystem, they interact with new resources and develop new strategies for digestion: new enzymes, new acids, new antibiotics. They are constantly learning and adapting not just from a single contact point but from countless points. During this process, the mycelium becomes more knowledgeable; its experience becomes ingrained into the very DNA of the network. Because these are self-learning membranes and highly adaptive, their remediative potential has nearly limitless applications.

I believe that the internet was invented as an inevitable consequence of a previously proven successful evolutionary model—mycelium. Mycelium is a network, our brain is a network, and our computer information-sharing systems are networks. Networks are valuable because of their resilience to change, their ability to adapt to change, and their ability to experience change. This is important because the more experiences you have, the broader your basis for making intelligent decisions. It's true for humans, it's true for mycelium, and now it's becoming true with hardware as artificial intelligence (AI) evolves into "learning machines" designed to "think" on their own as they "interpret" countless algorithmic processes. Such machines are now able to determine, for example, whether songs or texts are happy or sad.[2] They're writing musical compositions and having conversations that are indistinguishable from conversations between humans.[3]

I envision a future where we are able to build synaptic junctions between our consciousness and nature, and this is when I think we will understand that the whole is much greater than the sum of its parts. A truly holistic consciousness must include a sense of the consciousness of all the organisms on Earth, and even a sense of something transcendent, what some would call God. It's all one, ultimately, but we have broken the full experience of life into all these disconnected parts.

TIME TO CHANGE THE GAME

The earth is in crisis, but I still believe in the power of beauty, gratitude, and forgiveness. When you are generous to people, they tend to be more generous back. When we are generous with nature, it responds in kind. Nature is designed to incentivize our basic goodness.

There's a new blending of capitalism, spirituality, and environmental philosophy that is starting to come together. If we can help show future generations a path that is worthy of pursuit, however difficult it may be at times, we can set the stage for enlightened thinking that leads to outcomes both commercially successful and healthy for planetary sustainability. And you need both for the scale of change we need, because in the current reality, that's where the resistance shows up: "Change will be too expensive. It will threaten our quality of life." But of course, the opposite is true if we don't alter our course; if we keep going the way we're going, our quality of life will become unsustainable.

Nevertheless, I think this is the crack in the cosmic egg, the dawning of a new awareness. We're on the cusp of a mycological revolution where an emerging ecology of consciousness is rooted in practical solutions we can analyze, test, and verify. The public has been under ceaseless assault by negative messages about the environment, the collapse of societies, disease, terrorism, and poverty. There's a tremendous unease out there and it can get pretty depressing. People are increasingly desperate and want to know what can be done. And that's why I'm optimistic, because my fungal allies have taught me that they have solutions literally underfoot that can respond to challenges very, very quickly.

Extinction events are chasing our tail. If we don't get in front of them, they will surpass and eclipse us, and then the game is over. Do we have the foresight, the knowledge, and the intelligence to form our future in a sustainable way by not dismantling the very foundations that make life possible? I believe we do. I believe that goodness and wisdom will ultimately triumph—with the help of the third kingdom.

NOTES

1. Kevin Loria, "How Psychedelics Like Psilocybin and LSD Actually Change the Way People See the World," BusinessInsider.com, February 26, 2017.
2. Terence Mills, "Machine Learning vs. Artificial Intelligence: How Are They Different?" Forbes.com, July 11, 2008.
3. Pranav Dar, "Google Is Making Music with Machine Learning—and Has Released the Code on GitHub," AnalyticsVidhya.com, March 14, 2018; Yongdong Wang, "Your Next New Best Friend Might Be a Robot," Nautilus.us, February 14, 2016.

ABOVE LEFT *Rendering of a prehistoric fungi-dominated ecosystem.* **ABOVE RIGHT** *Rendering of an ice age.* **OPPOSITE TOP** Marasmiellus. **OPPOSITE BOTTOM** Phallus indusiatus, *an edible and choice stinkhorn; that is, until it matures and becomes very stinky.*

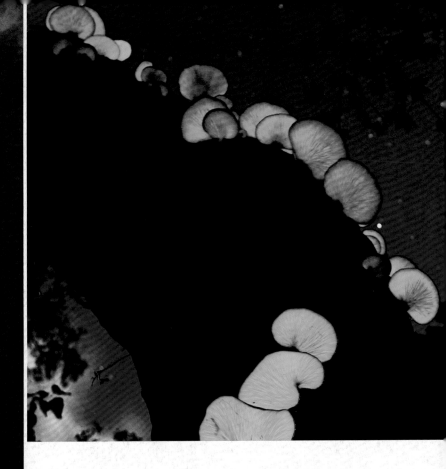

"Anyone who's had a powerful psychedelic experience in a country where it's illegal to have one faces a dilemma—the risk of prosecution—because something precious and profound, a great personal gift, is not understood by the culture at large. One response is to get angry and fight that. Another is to say that we need to do a better job explaining the phenomenon. When people understand it, the potential for accommodation and respect opens up."

—BOB JESSE, FOUNDER OF THE COUNCIL ON SPIRITUAL PRACTICES (CSP)

LEFT AND ABOVE Omphalotus *sp*. *These bioluminescent mushrooms glow because they contain molecules called luciferins, which upon oxidation produce a virtually heatless light.* FOLLOWING PAGES FROM TOP LEFT TO BOTTOM RIGHT Phlogiotis helvelloides; Nidula *sp.*; Pholiota squarrosa; Ramaria *sp.*; Gyromitra esculenta; Humaria hemisphaerica; Helvella macropus; Auricularia delicata; Pseudohydnum gelatinosum; Ascocoryne sarcoides; Stemonitis axifera; Fistulina hepatica. *(All images on page © Taylor Lockwood.)*

This world of ours is always changing . . .
not for the better, or the worse, but for life.
If the storms come and the water rises . . .
If fire scorches the land . . .
Or darkness descends . . .
We will be here, working. As we always have.
Extending the network, building community,
restoring balance, little by little.
One connection at a time.
It may take a million years, or a hundred million.
But we will still be here.

AFTERWORD

LOUIE SCHWARTZBERG

LOUIE SCHWARTZBERG is an award-winning producer, director, and cinematographer whose notable career spans more than three decades. Known for his time-lapse, high-speed, and macro cinematography, Louie's films are stories that celebrate life, make the invisible visible, and reveal the mysteries and wisdom of nature, people, and places.

The journey toward making the film *Fantastic Fungi* started almost thirteen years ago when I heard Paul Stamets give one of his first presentations at the Bioneers Conference. By that time, I had already been seduced by the sensual beauty of flowers and had been time-lapsing them twenty-four hours a day, seven days a week, nonstop for over three decades. Part of that effort included time-lapsing mushrooms. After Paul's presentation, I showed him some of the time-lapse videos of mushrooms I had on my laptop. In that moment, the mycelium network successfully made its intentional connection. That is what the mycelium network does: It connects living beings so life can flourish and so we can live in harmony with the earth.

My passion for capturing imagery that inspires wonder and awe, and for capturing subjects that are too slow, too fast, too small, or too vast for the naked eye to see, is what led me to filmmaking. I love taking audiences through portals of time and scale. These immersive experiences are transcendent and broaden our worldview.

For a while, I thought that my obsession with capturing the beauty of a flower's infinitesimal movement was enough of a reason for me to keep cameras filming around the clock. But then I learned about colony collapse disorder, the scientific name for the mass death of bee colonies, and I knew I could not tell the story of flowers without telling the story of the bees' decline. Bees coevolved with flowers—a love affair that has been going on for over fifty million years—and we can't let that relationship unravel. If the bees go, we go. All of the foods we need to stay healthy—including fruits, vegetables, nuts, seeds, berries, and some grains—come from pollinating plants. Without the evolution of the flower, warm-blooded mammals would never have evolved. Before the appearance of flowers, Earth was a boring and drab place, mostly green. The cold-blooded herbivorous reptiles roaming the planet needed to eat huge quantities of leaves to stay alive.

So . . . I made a film about pollination called *Wings of Life*. It is a Disneynature feature film narrated by Meryl Streep, who is the voice of a flower who seduces bees, bats, butterflies, and hummingbirds. I learned through that filmmaking process that pollination is a keystone event, a magical intersection and interaction between the animal and plant kingdoms that happens billions of times each day. Without this event, life on Earth would be radically different.

I love exploring the big questions and diving into the mysteries of life. So, I thought, if plants are the only land-based life that can convert the sun's energy into food, which is critical for our own and other creatures' survival, what do plants need in order to survive? The answer is soil.

Now, let's take it further. Where does soil come from? What can break down organic matter, including minerals from rock, to make soil? The answer surprised me. It is the largest organism on the planet, one that is everywhere, on every continent, under our feet, and inside our bodies: fungi! So when I heard Paul's presentation, I instantly knew that I was going to take a deep dive into the world of fungi and that I needed to make a film about this foundational aspect of life, which most of us know nothing about.

It has been a truly fantastic journey. Learning how mushrooms can feed us, heal us, clean up environmental pollution—including the atmosphere—and shift our consciousness has changed my life. I interviewed scientific experts, who shared their knowledge on all these topics, and featured those leading experts in the film in a format in which they could present those science-based facts to the general public.

Making the film was a great learning experience. Yet, it goes beyond education on some astounding levels. The beauty of the time-lapse mushrooms dancing as they emerge from their underground world both entertains and inspires. The journey through the underground mycelium network—which mirrors the networks of our own nervous and circulatory systems as well as of the internet and of galaxies in space—takes my breath away and has the power to bring audiences to tears of joy. I never knew that fungi, along with their plant partners, could be the greatest and fastest natural solution to climate change. The Mother Tree concept and underground internet idea—pioneered by Paul Stamets and Suzanne Simard—was the foundational science for concepts filmmaker James Cameron used in *Avatar*. Without that spiritual core, *Avatar* would never have become one of the top box office films of all time.

Yet, my biggest takeaway was learning that the mycelium network creates connections between plants and trees that enable ecosystems to flourish as symbiotic communities. Nothing lives alone in nature, and communities are more likely to survive than individuals. What a beautiful inspirational model for how human beings might live: In a shared economy based not on greed but on nurturing relationships and mutual cooperation.

One of the many wonder*full* aspects of making the *Fantastic Fungi* film was the collaboration between artists and scientists. There is an intrinsic relationship between science and art, and both foster an incredible sense of wonder. The underground mycelium sequences, which were animated by artists, used scanning electron microscopy images as references. The time-lapse sequences of growing mushrooms used the art of lighting and cinematography. And, bringing this arc full circle, I implored Paul Stamets, leading mycologist and brilliant inventor, to come up with a solution to save the bees. Paul embraced that challenge like a true Jedi Knight and developed mushroom extracts that have been proven in the field to save bees from the viral infections that are the major identified causes of their decline. That solution could save the world's food supply.

We are now living in times of environmental breakdown and technological breakthroughs. We have barely scratched the surface of the fungal genome, which may prove to be one of our most important partners going forward. I am hopeful about the future, because answers to our greatest problems might be literally right under our feet. We need to open our eyes to nature's intelligent design and our hearts to nature's wonders. If we can embrace our fungal partners and ancestors, we can change the fate of our planet from mass extinctions to flourishing environments, and in so doing, foster communities—both for us and for future generations—that celebrate and honor life in its many forms.

ACKNOWLEDGMENTS

LOUIE SCHWARTZBERG

I want to express my sincerest gratitude for these folks who encouraged and supported the film from its earliest days, without whom the film and this book could not have been realized: my producing partners: Lyn Lear and Elease Lui Stemp.

Leading the charge, champions and fellow eco-warriors: Regina Scully, Margaret Bear, Elizabeth Parker, Anna Getty, Jena King, Shannon O'Leary Joy, Bill and Laurie Benenson, Carol Newell, Geralyn Dreyfous, Kenny Ausubel, Melony Lewis, Susan Rockefeller, Norman Lear, Cindy Horn, and our Kickstarter network.

Making a film is no easy feat! My mushroom cap goes off to those who went through the trenches with me, metaphorically and literally, to get this courageous, groundbreaking, rebellious, and beautiful film out to the world: Annie Wilkes, Kevin Klauber, Courtney Mahler, Alex Falk, Mark Monroe, Adam Peters, Wylie Stateman, Sara Ramo, Barnaby Steel, Robin Aristorenas, and Victoria Mesesan.

And an extra special thanks to the fungi tribe who inspired and guided the film: Paul Stamets, Suzanne Simard, Eugenia Bone, Michael Pollan, Andrew Weil, Tony Bossis, Bill Richards, Roland Griffiths, Bill Linton, Bob Jesse, Charles Grob, Dennis McKenna, Jay Harman, Alex and Allyson Grey, Gary Lincoff, and the vibrant Telluride Mushroom Festival.

Special thanks to Brie Larson, who embodied the voice of the mushrooms, and Raoul Goff, our publisher, who creates the most beautiful and soul-inspiring books with Insight Editions, Earth Aware Editions, and Mandala Publishing.

ABOVE Coprinellus *sp.* OPPOSITE FROM TOP LEFT TO BOTTOM RIGHT *Unidentified mushroom species;* Cortinarius *sp.;* Cortinarius *sp.;* Arnillaria mellea.

INDEX

A
Agaricus bisporus mushroom, 84, 90
agarikon (*Laricifomes officinalis* mushroom, 113-114
almond pollination study, 51
Alzheimer's disease, 76, 130
amadou (*Fomes fomentarius*), 51
Amanita muscaria (fly agaric), 12
Amazon Mycorenewal Project (now CoRenewal), 59
anorexia nervosa, 130
ayahuasca (plant-based spiritual medicine), 143

B
Barrett, Frederick, 130
Bastyr University-University of Minnesota Medical School study, 78
Bavarian Beer Purity Act Law (*Reinheitsgebot*) [1516], 12
Beckley Foundation (UK), 151
bee industry, 15
Big Pharma, 122
biodiveristy, 23, 50, 66, 116, 123
biomimetic solutions, 27-29, 31
biomimicy, 27-29, 31
bird (H5N1) flu, 114, 116
Black Queen Cell virus, 51
Blair, Katrina, 107
Bolt Threads, 33
Bone, Eugenia, 81, 87
Bossis, Anthony, 157, 158
the brain, 149-150, 154, 163
breast cancer study, 78
Brooklyn's Prospect Park, 94
Bunyard, Britt, 97

C
carbon dioxide (CO_2), 20, 23
Changing Our Minds Psychedelic Sacraments and the New Psychotherapy (Lattin), 151
Chile, 42-43
Chilean Ministry of the Environment, 42
Chinese medicine, 51, 76-77, 85
Clemson University, 62
climate change, 20, 31, 69
community labs, 57
COMPASS Pathways, 130
Connecticut-Westchester Mycological Association (COMA), 95
CoRenewal (Amazon Mycorenewal Project), 59
Cosimano, Mary, 137
Cotter, Tradd, 61, 105
Council on Spiritual Practices (CSP), 157, 159
Cozzi, Nicholas V., 125

D
Darwin, Charles, 21
death and end-of-life studies, 143-148, 163
deformed wing virus (DWV), 49, 51
dementia, 76
Department of Defense BioDefense program, 15
depression studies, 163
desert truffles, 98

E
Earth: mycelium networks as having the answers to saving the, 42, 43; mycelium role in forest recycling and renewal of life on, 19-20; pollination's role in health of, 15, 49-52
Eastern honeybee (*Apis cerana*), 49
eating mushrooms, 84-85, 88-89, 107
ecosystems, 35-36, 50, 65-66
ecotoxicological test, 55-56
Ecovative, 33
education. *See* myco-literacy and activism
end-of-life studies, 143-148, 163
endophytic fungi, 83
Enjoying Wild Mushrooms (Lincoff), 95
environmental damage: biomimetic solutions for, 28-29, 31; climate change, 20, 31, 69; growing awareness and commitment to solutions for, 27-29; Hawaii's sunscreen ban to prevent further, 28; loss of honeybees due to, 15, 49-52; mycoremediation to help restore, 55-59. *See also* pollution
environmental health, 42-43, 50
evolution, 20-21, 66, 69, 75
extinction rate, 27

F
Fadiman, James, 151
Fantastic Fungi (film), 15, 174, 175
Farlow Herbarium (Harvard University), 43
Fish and Wildlife Department, 31
Fleming, Alexander, 115
flora, fauna, and Fungi (three F's), 42
Fomes fomentarius (tinder fungus), 33
Fomitopsis officinalis, 12
Food and Drug Administration (FDA), 115, 121, 127, 130
forests, 19-21, 23, 35-36, 83-84
fungal cleaners list, 59

fungal intelligence, 32, 165
fungal parasites, 84
fungi: agricultural environments and role of, 82-83; antimicrobial properties of, 12; connecting plants to animals, 42, 43, 61, 75, 87; health of the earth connected to health of, 41, 43; mycorrhizal, 35-36; saprobic, mutualists, and endophytic, 83; spores, 37, 85-86; understanding the nature of, 82. *See also* mushrooms; mycelium networks
fungi extracts, 15, 51
Fungi Foundation, 42
Fungi Magazine, 97
Fungi Perfecti, 51
fungi research: community labs and citizen, 57; creating international partnerships in, 43; developing certification and field guides for, 42; expansion of, 66, 69; findings on the ecology of mycorrhizal fungus networks, 35-36; on fungi extracts used to rescue honeybees, 51; the long-term mission of, 65-66. *See also* medicine; mycelium networks; psilocybin ("magic mushrooms")
Furci, Giuliana, 41

G

Garcia-Romeu, Albert, 130
Getty, Adele, 140
Goodall, Jane, 42
green chemistry, 31
greenhouse gasses, 20
Grey, Alex, 132
Griffiths, Roland, 157
Grifola frondosa mushroom, 84
Grob, Charles, 147
Grof, Stanislav, 143
growing mushrooms, 105-106, 110
Gulf of Mexico oil spill, 55

H

Haereticus Environmental Laboratory, 28
Haiti, 61, 62
Harman, Jay, 27, 28
Hawaii's sunscreen ban, 28
health benefits: of agarikon (*Laricifomes officinalis*) mushroom, 113, 114; Chinese medicine use of mushrooms for the, 51, 76-77, 85; the immune system and mushrooms, 113-115; of medicinal mushrooms, 15, 76-79, 85-86; of mushrooms for honeybees, 15, 49-52
Heffter Research Institute, 159
Heim, Roger, 125
High Times magazine, 63
Hofmann, Albert, 125
Holstrom, Kris, 105
honeybees, 15, 49-52

How to Change Your Mind (Pollan), 97, 121

I

immune system, 85, 115-116
Imperial College London, 163
Industrial Revolution, 27, 31
International Union for Conservation of Nature (IUCN), 42
internet metaphor, 35

J

Jamaican mushroom industry, 62
Jesse, Robert, 129, 157, 158
Jobs, Steve, 151
Johns Hopkins studies, 122, 130, 138-139, 146, 157
Johnson, Matthew, 130

K

Kavaler, Lucy, 63

L

Lake Sinai virus, 51
Lattin, Don, 151
Leucaena leucocephala (white leaf tree or river tamarind), 20
Lincoff, Gary, 93, 95, 97
Lion's mane mushroom, 76
Local Wildlife: Turtle Lake Refuge Recipes for Living Deep (Blair), 107
Louv, Richard, 69
LSD, 121, 143, 151, 153
LSD Psychotherapy: The Healing Potential of Psychedelic Medicine (Grof), 143

M

"magic mushrooms." *See* psilocybin ("magic mushrooms")
Mazatec mushroom ritual, 125
McCoy, Peter, 45
McKenna, Dennis, 153
McKenna, Terence, 153
medicinal mushrooms: antiviral properties of mushrooms, 114-115; ayahuasca (plant-based spiritual medicine), 143; health benefits of, 15, 76-79, 85-86; list of health benefits and varieties of, 79. *See also* psilocybin ("magic mushrooms")
medicine: breast cancer study, 78; Chinese medicine, 51, 76-77, 85; discovery of penicillin, 85, 115; end-of-life studies, 143-148, 163; immune system and mushrooms, 85, 115-116; innovations using mycelium networks, 32; psychedelic therapy, 121-128, 130, 146-147, 149-150, 163-164. *See also* fungi research; immune system
meditation, 160
Metarhizium, 51
Money, Nik, 37
Mother Trees, 20-21, 175

mushroom community, 57-58, 63, 94, 100-101
The Mushroom Cultivator (Stamets), 63
Mushroom Mountain (farm and research facility), 61, 106
mushrooms: the amazing history and benefits of, 11-15, 93-94, 103; community lab work sorting, 57; education for cultivating myco literacy on, 45; foodie treat and recipes, 84-85, 88-89, 107; as fruiting bodies of fungi, 82; growing your own, 105-106, 110; health benefits and medical uses of, 15, 76-79, 85-86, 113-116; Jamaican industry of, 62; as part of a hidden network of mycelium, 19-21; radical mycology on theory and practice of using, 45; wild mushrooming and urban foraging, 94, 101. *See also* fungi
Mushrooms, Molds, and Miracles (Kavaler), 63
Mushrooms, Russia, and History (Wasson and Wasson), 125
mutualists, mycorrhizal fungi, 83
mycelial intelligence, 32, 165
mycelium activism, 57-58, 63
mycelium extracts, 15, 51
mycelium networks: as having the answers to saving our planet, 42, 43; mushrooms as part of hidden, 19-21; mycelium webbing found in soils, 61-62; "mycomimicry" for building a sustainable society, 27; Mylo material made from cells of, 33; as part of the forest recycling and renewal of life, 19-20; plants supported by, 32, 35, 42, 43, 65; representing ecological remedies, 65-66. *See also* fungi; fungi research
myco-literacy and activism, 43, 45, 47
mycoheterotrophs, 35-36
a mycologist's journey, 109-111
Mycologos school, 45
mycology, 35-36, 66, 69, 86-87
mycomimicry, 27-29, 31
mycophiles, 57, 58, 63, 81
mycophilia, 86
mycoremediation, 55-59
mycorrhizal fungus networks, 20-21, 35-36, 83-84
MycoSymbiotics, 110
MycoWorks, 33
Mylo material (Bolt Threads), 33

N

National Audubon Society Field Guide to North American Mushrooms (Lincoff), 95
National Environment Bill (Chile), 42
National Institutes of Health laboratories, 113
Native American Church (NAC), 143
natural selection, 66
nature: as blueprint for building a sustainable society, 27-29; historic extinction rate in, 27; the interconnectedness of, 69; Mother Trees partnership with mycorrhizal fungus in, 20-21; species dealing with UVA or UVB in, 29, 83; survival of the fittest, competition, and healing in, 20-21
New York Botanical Garden, 95
New York City mushroom foraging, 94
New York Mycological Society, 94, 95
Newtonian physics, 27
nonmelanoma skin cancer, 28
North American Mycological Association (NAMA), 95
Nutt, David, 163

O

octinoxate, 28
Okumenisches Haus der Stille (House of Silence), 160
old-growth forests, 23
Olson, Leif, 105
Ophiocordycep sinensis (caterpillar mushroom), 85
Ornilux, 29
Otogawa, Kobun Chino, 160
oxybenzone, 28

P

Padilla-Brown, William, 109
Palmers, Vanja, 160
penicillium mold, 85, 115
Penicillium, 12
Penne al Funghi with Arugula recipe, 88
Pfister, Don, 43
photosynthesis, 20, 35-36
plants: cooperation between trees and, 21; fungal parasites prey on, 84; fungi connecting animals to, 42, 43, 61, 75, 87; mycelium support of, 32, 35, 42, 43, 65; mycoheterortrophs, 35-36; photosynthesizing carbon dioxide (CO_2), 20
"plug spawns," 105-106
Pollan, Michael, 97, 121
pollination, 15, 49-52, 174
pollution, 20, 31, 49, 56. *See also* environmental damage
polyaromatic hydrocarbons, 55
polyphenols, 113
polypore mushroom extracts, 15
polypores mushrooms, 76
Psilocybin: Magic Mushroom Growers Guide (McKenna and McKenna), 153
psilocybin ("magic mushrooms"): changing the game of medicine, 163-167; creating sacred spaces for exploring, 137; Johns Hopkins session transcripts on, 138-139; LSD, 143, 151, 153; meditation and, 160; psychedelic experiences with, 12, 15, 97, 122, 125, 129-130, 131, 140, 169; psychedelic therapy applications of, 121-128, 130, 146-147, 149-150; stoned ape hypothesis, 154-155. *See also* fungi research; medicinal mushrooms
psychedelic experiences, 12, 15, 97, 122, 125, 129-130, 131, 140, 169

psychedelic therapy, 121-128, 130, 146-147, 149-150, 153-154
Psychiatric Research Institute (Prague), 143

R
radical mycology, 45
Radical Mycology Mycelial Network, 47
Red List of Threatened Species Criteria (IUCN), 42
reishi mushrooms (*Ganoderma lucidum*), 51
religious experiences, 143, 157-159
research. *See* fungi research; medicine
restoration projects, 55-59
Reyes, Daniel, 105
Richards, William, 129, 151, 157, 158
Richardson, Allan, 125
Ross, Philip, 33
Ross, Stephen, 143, 157
Royal Botanic Gardens (Kew, England), 43

S
Sabina, María, 125
saprobic fungi, 83
Schaechter, Moselio, 87
Schwartzberg, Louie, 15, 174
"Seeking the Magic Mushroom" (Wasson), 125
serotonin, 126
Shavit, Elinoar, 98, 103
Sheldrake, Merlin, 35
Sheppard, Steve, 49
Simard, Suzanne, 19, 175
soil: mycelium webbing found in, 61-62; mycorrhizal fungus networks impact on, 20-21, 35-36, 83-84; work to restore health of, 55-59
Soma: Divine Mushroom of Immortality (Wasson), 125
spiderwebs, 29
spores, 37, 85-86
Stamets, Paul, 11, 23, 42, 51, 63, 65, 79, 87, 97, 100, 105, 113, 114, 163, 174, 175
Steindl-Rast, David, 160
stoned ape hypothesis, 154-155
stormwater runoff, 56
Streep, Meryl, 174
stuttering cure story, 164
sunscreen ban (Hawaii), 28
survival of the fittest, 20-21, 66
sustainable society, 27-29, 66

T
Taxomyces andreanae, 12
tecovirimat (TPOXX), 115
Telluride Mushroom Festival, 97, 105
Texas Commission on Environmental Quality (TCEQ), 56
third kingdom. *See* mycelium networks

Tibetan ghost moth larvae, 85-86
timeline of mushrooms in civilization, 103
Tirmania nivea desert truffle, 98
Tomten Farms, 105
Toynbee, Arnold, 132
truffles, 84-85, 98
tuberculosis (TB), 114
Turkey tail (*Trametes versicolor*), 76

U
ultraviolet A (UVA), 28, 29
ultraviolet B (UVB), 28, 29
União do Vegetal (UDV) religion, 143
United States Army Medical Research Institute of Infectious Diseases, 114
United States Department of Agriculture, 15
University of Concepción (Chile), 42
urban mushroom foraging, 94
USDA, 82, 90
Usona Institute, 130
UVA (or UVB), 28, 29
UW-Madison Phase 1 study, 127

V
Varroa mite (*Varroa destructor*), 49, 50
Vollenweider, Franz, 149

W
Washington State University, 15
Wasson, Robert Gordon, 125
Wasson, Valentina Pavlovna, 125
water contaminants, 56
Weil, Andrew, 75, 105, 155
Western honeybees (*Apis mellifera*), 49
Wild Mushroom Soup recipe, 89
Wild Mushroom Telluride (later Telluride Mushroom Festival), 97
wild mushrooming, 94, 101, 107
The Wild Wisdom of Weeds: 13 Essential Plants for Human Survival (Blair), 107
Wings of Life (film), 15, 174
The Wondrous Mushroom: Mycolatry (Wasson), 125
wood-chip substrates, 106
Wood Wide Web, 35-36
World Health Organization, 61, 115

Z
Zvnder, 33

EARTH AWARE

An Imprint of MandalaEarth
PO Box 3088
San Rafael, CA 94912
www.MandalaEarth.com

Find us on Facebook: www.facebook.com/MandalaEarth
Follow us on Twitter: @mandalaearth

Photographs on pages 2-3, 24-25, 53, 156, 182-183 © Art Wolfe
Photographs on pages 10, 13, 38-39, 40, 44, 70-71, 112, 152, 158, 159, 176 © Ian Shive
Photographs on pages 58, 114, 122, 124, 127, 128, 144, 147, 152 © Paul Stamets
Photographs on pages 10, 13, 14, 18, 29, 54, 56, 64, 102, 116, 126, 142, 170-171 © Taylor Lockwood
Photographs on page 33 © Bolt Threads
Photographs on pages 133 © Alex Grey
All photographs not specified above copyright © Moving Art
Introduction copyright © 2019 Paul Stamets
Afterword copyright © 2019 Louie Schwartzberg
Text copyright © 2019 Earth Aware Editions & Moving Art

Text written by Matthew Gilbert; all essays were based on interview transcripts and were reviewed or revised by interviewees when possible.

Published by Earth Aware Editions, San Rafael, California, in 2019. No part of this book may be reproduced in any form without written permission from the publisher.

Library of Congress Cataloging-in-Publication Data available.

ISBN: 978-1-68383-704-6

PUBLISHER: Raoul Goff
CREATIVE DIRECTOR: Chrissy Kwasnik
DESIGNER: Amy DeGrote
SENIOR EDITOR: Kelly Reed
EDITOR: Courtney Andersson
ASSOCIATE MANAGING EDITOR: Lauren LePera
ASSISTANT EDITOR: Tessa Murphy
SENIOR PRODUCTION EDITOR: Rachel Anderson
SENIOR PRODUCTION MANAGER: Greg Steffen

Earth Aware Editions wishes to thank Phillip Jones and Kate Jerome for their editorial contributions and Jon Glick and Katherine Yao for their design contributions.

Although the authors, editors, and publishers have made every effort to ensure that the information in this book was correct at press time, the author and publisher do not assume and hereby disclaim any liability to any party for any loss, damage, or disruption caused by errors or omissions, whether such errors or omissions result from negligence, accident, or any other cause.

This book is not intended as a substitute for the medical advice of physicians. The reader should consult a physician in matters relating to his/her health.

Every effort has been made to correctly identify mushrooms, but please consult a mushroom identification guide to ensure accurate identification.

Manufactured in China by Insight Editions

20 19 18 17 16 15 14 13 12

PAGES 182-183 *Parasol mushroom* (Chlorophyllum rachodes).

TOP *Unidentified mushroom species.*

ABOVE *Two red fruit bodies of a slime mold or myxomycete* (Physarum roseum).

IN MEMORIAM

Patricia North Stamets, ninety-three years old, mother of Paul Stamets, died during the final stages of creating *Fantastic Fungi*. Her support and dedication constantly inspired Paul to follow his mycelial path, and she instilled in him the desire to carry forth the spirit of goodness.

Earth Aware Editions, in association with Roots of Peace, will plant two trees for each tree used in the manufacturing of this book. Roots of Peace is an internationally renowned humanitarian organization dedicated to eradicating land mines worldwide and converting war-torn lands into productive farms and wildlife habitats. Roots of Peace will plant two million fruit and nut trees in Afghanistan and provide farmers there with the skills and support necessary for sustainable land use.